Lecture Notes in Mathematics

Vols. 1-183 are also available. For further information, please contact your book-seller or Springer-Verlag.

Vol. 184: Symposium on Several Complex Variables, Park City, Utah, 1970. Edited by R. M. Brooks. V, 234 pages. 1971. DM 20,-

Vol. 185: Several Complex Variables II, Maryland 1970. Edited by J. Horváth. III, 287 pages. 1971. DM 24,-

Vol. 186: Recent Trends in Graph Theory. Edited by M. Capobianco/ J. B. Frechen/M. Krolik. VI, 219 pages. 1971. DM 18,-

Vol. 187: H. S. Shapiro, Topics in Approximation Theory. VIII, 275 pages. 1971. DM 22,-

Vol. 188: Symposium on Semantics of Algorithmic Languages. Edited by E. Engeler. VI, 372 pages. 1971. DM 26,-

Vol. 189: A. Weil, Dirichlet Series and Automorphic Forms. V, 164 pages. 1971. DM 16,-

Vol. 190: Martingales. A Report on a Meeting at Oberwolfach, May 17-23, 1970. Edited by H. Dinges. V, 75 pages. 1971. DM 16,-

Vol. 191: Séminaire de Probabilités V. Edited by P. A. Meyer. IV, 372 pages. 1971. DM 26,-

Vol. 192: Proceedings of Liverpool Singularities - Symposium I. Edited by C. T. C. Wall. V, 319 pages. 1971. DM 24,-

Vol. 193: Symposium on the Theory of Numerical Analysis. Edited by J. Ll. Morris. VI, 152 pages. 1971. DM 16,-

Vol. 194: M. Berger, P. Gauduchon et E. Mazet. Le Spectre d'une Variété Riemannienne. VII, 251 pages. 1971. DM 22,-

Vol. 195: Reports of the Midwest Category Seminar V. Edited by J. W. Gray and S. Mac Lane. III, 255 pages. 1971. DM 22,-

Vol. 196: H-spaces - Neuchâtel (Suisse)- Août 1970. Edited by F. Sigrist. V, 156 pages. 1971. DM 16,-

Vol. 197: Manifolds - Amsterdam 1970. Edited by N. H. Kuiper. V, 231 pages. 1971. DM 20,-

Vol. 198: M. Hervé, Analytic and Plurisubharmonic Functions in Finite and Infinite Dimensional Spaces. VI, 90 pages. 1971. DM 16,-

Vol. 199: Ch. J. Mozzochi, On the Pointwise Convergence of Fourier Series. VII, 87 pages. 1971. DM 16,-

Vol. 200: U. Neri, Singular Integrals. VII, 272 pages. 1971. DM 22,-

Vol. 201: J. H. van Lint, Coding Theory. VII, 136 pages. 1971. DM 16,-

Vol. 202: J. Benedetto, Harmonic Analysis in Totally Disconnected Sets. VIII, 261 pages. 1971. DM 22,-

Vol. 203: D. Knutson, Algebraic Spaces. VI, 261 pages. 1971. DM 22,-

Vol. 204: A. Zygmund, Intégrales Singulières. IV, 53 pages. 1971. DM 16,-

Vol. 205: Séminaire Pierre Lelong (Analyse) Année 1970. VI, 243 pages. 1971. DM 20,-

Vol. 206: Symposium on Differential Equations and Dynamical Systems. Edited by D. Chillingworth. XI, 173 pages. 1971. DM 16,-

Vol. 207: L. Bernstein, The Jacobi-Perron Algorithm - Its Theory and Application. IV, 161 pages. 1971. DM 16,-

Vol. 208: A. Grothendieck and J. P. Murre, The Tame Fundamental Group of a Formal Neighbourhood of a Divisor with Normal Crossings on a Scheme. VIII, 133 pages. 1971. DM 16,-

Vol. 209: Proceedings of Liverpool Singularities Symposium II. Edited by C. T. C. Wall. V, 280 pages. 1971. DM 22,-

Vol: 210: M. Eichler, Projective Varieties and Modular Forms. III, 118 pages. 1971. DM 16,-

Vol. 211: Théorie des Matroïdes. Edité par C. P. Bruter. III, 108 pages. 1971. DM 16,-

Vol. 212: B. Scarpellini, Proof Theory and Intuitionistic Systems. VII, 291 pages. 1971. DM 24,-

Vol. 213: H. Hogbe-Nlend, Théorie des Bornologies et Applications. V, 168 pages. 1971. DM 18,-

Vol. 214: M. Smorodinsky, Ergodic Theory, Entropy. V, 64 pages. 1971. DM 16,-

Vol. 215: P. Antonelli, D. Burghelea and P. J. Kahn, The Concordance-Homotopy Groups of Geometric Automorphism Groups. X, 140 pages. 1971. DM 16,-

Vol. 216: H. Maaß, Siegel's Modular Forms and Dirichlet Series. VII, 328 pages. 1971. DM 20,-

Vol. 217: T. J. Jech, Lectures in Set Theory with Particular Emphasis on the Method of Forcing. V, 137 pages. 1971. DM 16,-

Vol. 218: C. P. Schnorr, Zufälligkeit und Wahrscheinlichkeit. IV, 212 Seiten. 1971. DM 20,-

Vol. 219: N. L. Alling and N. Greenleaf, Foundations of the Theory of Klein Surfaces. IX, 117 pages. 1971. DM 16,-

Vol. 220: W. A. Coppel, Disconjugacy. V, 148 pages. 1971. DM 16,-

Vol. 221: P. Gabriel und F. Ulmer, Lokal präsentierbare Kategorien. V, 200 Seiten. 1971. DM 18,-

Vol. 222: C. Meghea, Compactification des Espaces Harmoniques. III, 108 pages. 1971. DM 16,-

Vol. 223: U. Felgner, Models of ZF-Set Theory. VI, 173 pages. 1971. DM 16,-

Vol. 224: Revètements Etales et Groupe Fondamental. (SGA 1). Dirigé par A. Grothendieck XXII, 447 pages. 1971. DM 30,-

Vol. 225: Théorie des Intersections et Théorème de Riemann-Roch. (SGA 6). Dirigé par P. Berthelot, A. Grothendieck et L. Illusie. XII, 700 pages. 1971. DM 40,-

Vol. 226: Seminar on Potential Theory, II. Edited by H. Bauer. IV, 170 pages. 1971. DM 18,-

Vol. 227: H. L. Montgomery, Topics in Multiplicative Number Theory. IX, 178 pages. 1971. DM 18,-

Vol. 228: Conference on Applications of Numerical Analysis. Edited by J. Ll. Morris. X, 358 pages. 1971. DM 26,-

Vol. 229: J. Väisälä, Lectures on n-Dimensional Quasiconformal Mappings. XIV, 144 pages. 1971. DM 16,-

Vol. 230: L. Waelbroeck, Topological Vector Spaces and Algebras. VII, 158 pages. 1971. DM 16,-

Vol. 231: H. Reiter, L^1-Algebras and Segal Algebras. XI, 113 pages. 1971. DM 16,-

Vol. 232: T. H. Ganelius, Tauberian Remainder Theorems. VI, 75 pages. 1971. DM 16,-

Vol. 233: C. P. Tsokos and W. J. Padgett. Random Integral Equations with Applications to stochastic Systems. VII, 174 pages. 1971. DM 18,-

Vol. 234: A. Andreotti and W. Stoll. Analytic and Algebraic Dependence of Meromorphic Functions. III, 390 pages. 1971. DM 26,-

Vol. 235: Global Differentiable Dynamics. Edited by O. Hájek, A. J. Lohwater, and R. McCann. X, 140 pages. 1971. DM 16,-

Vol. 236: M. Barr, P. A. Grillet, and D. H. van Osdol. Exact Categories and Categories of Sheaves. VII, 239 pages. 1971. DM 20,-

Vol. 237: B. Stenström, Rings and Modules of Quotients. VII, 136 pages. 1971. DM 16,-

Vol. 238: Der kanonische Modul eines Cohen-Macaulay-Rings. Herausgegeben von Jürgen Herzog und Ernst Kunz. VI, 103 Seiten. 1971. DM 16,-

Vol. 239: L. Illusie, Complexe Cotangent et Déformations I. XV, 355 pages. 1971. DM 26,-

Vol. 240: A. Kerber, Representations of Permutation Groups I. VII, 192 pages. 1971. DM 18,-

Vol. 241: S. Kaneyuki, Homogeneous Bounded Domains and Siegel Domains. V, 89 pages. 1971. DM 16,-

Vol. 242: R. R. Coifman et G. Weiss, Analyse Harmonique Non-Commutative sur Certains Espaces. V, 160 pages. 1971. DM 16,-

Vol. 243: Japan-United States Seminar on Ordinary Differential and Functional Equations. Edited by M. Urabe. VIII, 332 pages. 1971. DM 26,-

Vol. 244: Séminaire Bourbaki - vol. 1970/71. Exposés 382-399. IV, 356 pages. 1971. DM 26,-

Vol. 245: D. E. Cohen, Groups of Cohomological Dimension One. V, 99 pages. 1972. DM 16,-

Vol. 246: Lectures on Rings and Modules. Tulane University Ring and Operator Theory Year, 1970-1971. Volume I. X, 661 pages. 1972. DM 40,-

continuation on page 135

Lecture Notes in Mathematics

Edited by A. Dold and B. Eckmann

405

Keith J. Devlin
Håvard Johnsbråten

The Souslin Problem

Springer-Verlag
Berlin · Heidelberg · New York 1974

Dr. Keith Devlin
Seminar für Logik und
Grundlagenforschung der
Philosophischen Fakultät
53 Bonn
Beringstraße 6
BRD

Dr. Havard Johnsbråten
Matematisk Institutt
Universitetet I Oslo, Blindern
Oslo 3/Norge

Library of Congress Cataloging in Publication Data

Devlin, Keith J
The Souslin problem.

(Lecture notes in mathematics, 405)
Bibliography: p.
1. Set theory. I. Johnsbraten, Havard, 1945-
joint author. II. Title. III. Series: Lecture
notes in mathematics (Berlin) 405.
QA3.L28 no. 405 [QA248] 510'.8s [511'.3] 74-17386

AMS Subject Classifications (1970): 02-02, 02K05, 02K25, 04-02, 04A30

ISBN 3-540-06860-0 Springer-Verlag Berlin · Heidelberg · New York
ISBN 0-387-06860-0 Springer-Verlag New York · Heidelberg · Berlin

Offsetdruck: Julius Beltz, Hemsbach/Bergstr.

Acknowledgements

The last five chapters of this book are based almost exclusively on a set of notes written by Devlin during the Summer of 1972. During this time, he was living with, and almost entirely supported by his wife's parents, Mr and Mrs H. Carey of Sidcup, Kent, to whom thanks should now be put on record.

The manuscript was typed by Mrs. R. Møller of the University of Oslo, for whose care and patience we owe a considerable debt.

CONTENTS

CORRECTION

An Israeli mathematician called Uri Avraham has pointed out that, in Chapter IX, the claim straddling pages 105 and 106 is false as stated. To be precise, given an Aronszajn tree $\mathring{T}$, we define, for C a certain closed unbounded subset of ω_1, a (generic) order-embedding of $\mathring{T}|C$ into $\mathbb{Q}$, and claim that this order-embedding can be extended to all of $\mathring{T}$. As we have set things up this is false (although a slight modification to the definition of $\tilde{T}_{\alpha+2}$ on page 105 would in fact make our claim true). What we should have said is not that our particular embedding extends to all of $\mathring{T}$, but that it easily gives rise to such an embedding. In fact, this is perhaps more easily seen (via Theorem II.5) in terms of our original definition of special Aronszajn (Page 15):

By means of our embedding of $\mathring{T}|C$ into $\mathbb{Q}$, we may write $\mathring{T}|C = \cup_{m<\omega} A_m$, where each A_m is an (uncountable) antichain of $\mathring{T}$. Let $\langle a^m_\alpha | \alpha<\omega_1 \rangle$ be a one-one enumeration of A_m, each m, and let $\langle c_\alpha | \alpha<\omega_1 \rangle$ be the monotone enumeration of C. For each $\alpha < \omega_1$, and for each $x \in \mathring{T}_{c_\alpha}$, let $S^x = \{y \in \mathring{T}|c_{\alpha+1} \mid x <_{\mathring{T}} y\}$. Since each S^x is countable, let $\langle s_n(x) | n< \omega \rangle$ enumerate it. For each $n,m < \omega$, the set $B_{n,m} = \{s_n(a^m_\alpha) | \alpha < \omega_1\}$ is clearly an antichain of $\mathring{T}$. But $\mathring{T} = (\cup_{n<\omega} A_n) \cup (\cup_{n,m<\omega} B_{n,m})$, so $\mathring{T}$ is special. QED.

Keith J. Devlin
Bonn, 14.9.74

Introduction

In 1920, the awakening of interest in Set Theory was heralded by the apprarence of a new journal in mathematics - Fundamenta Mathematicae. It is fitting that in the very first volume appeared the statement of a problem which was to play a prominent role in the development of that subject. Indeed, as with the continuum problem, research into the simple question there raised by M. Souslin was to have far-reaching ramifications in several branches of set theory, notably in constructibility theory and in the theory of forcing, the very two subjects whose existence stemmed from the work on the continuum problem. As with the continuum problem, Souslin's problem concerned the real numbers. However, instead of asking "how many" reals there are, Souslin asked whether a certain set of conditions uniquely characterized the real numbers. More precisely, Souslin's problem was this. Suppose we are given a dense linearly ordered set X, complete (in the sense of, say, Dedekind cuts) and with no end-points, and that X has the property that there is no uncountable collection of pairwise disjoint open intervals of X. Must then X be (isomorphic to) the real line, $\mathbb{R}$? (Problème 3, M. Souslin [Su 1].) It turned out that this problem was undecidable on the basis of the usual (i.e. Zermelo - Fraenkel) axioms of set theory. It is the purpose of this short book to provide the proofs of this undecidability and of the undecidability of the problem even when the continuum hypothesis is assumed. The vast majority of these results are due solely to one person, Ronald Jensen. (In fact, Chapter VI is essentially the only place where results appear which are not his.) For the most part, these results appear here for the first time in print.

The exposition falls naturally into two parts. In the five first chapters, we discuss the non-provability of Souslin's hypothesis (i.e. the assumption that Souslin's problem has a positive answer) in ZFC $\pm$ CH. This part which also contains (Chapter II) a basic discussion of the problem, should be fairly easy to read. The prerequisites are a reasonable (e.g. 1st year graduate level) aquaintance with the method of forcing and a knowledge of constructibility theory. In chapter 1 we sketch briefly the material we require. It is natural to include in this part some discussion of the notion of a Souslin line (i.e. a "continuum" which does possess the properties stated in Souslin's problem, but which is not isomorphic to $\mathbb{R}$), and Chapters IV and V provide just such a discussion. The second part - Chapters VI - X - is concerned entirely with the consistency of Souslin's hypothesis. Strictly speaking, the forcing theory outlined in Chapter I is all that is required for an understanding of the proofs, but some considerable aquaintance with this kind of material is really necessary. In particular, the proofs in this part tend to be extremely long and technical. We expect that chapters VIII - X will only be read in detail by the highly motivated (whatever that motivation may be !). These three chapters are certainly not the stuff of "first-year graduate courses" !

The book comes from two main sources. For the first part of the book, a set of lecture notes written by Jensen in Kiel in 1969. For the second part, a set of notes written by Devlin in Sidcup in 1972. These latter were based exclusively on a (much larger) set of notes written by Jensen some years earlier. For readers who have previously seen the original Jensen manuscript, let us say now that the proof of Con (ZFC + SH + CH) which we give here is essentially the same as the proof described there. The organisation of the proof is somewhat different, however, and the length is considerably reduced. Also (hopefully) we have found all the errors which the original manuscript contained.

We have tried to adopt the terminology and notation most common in current set-theoretical usage. We use 'iff' to denote 'if and only if' and use either QED or else ▌ to denote the end of a proof. Since this is not really a textbook (rather an account of Jensen's work) we have not included any exercises. The reader who completes the entire book will not need any further exercise !

We thank Thomas Jech for reading the last five chapters and suggesting some improvements.

Keith J. Devlin & Håvard Johnsbråten
Department of Mathematics
University of Oslo
June, 1974

Chapter I.

PRELIMINARIES

1. Set Theory

We shall be concerned entirely with the theory **ZFC** - Zermelo-Fraenkel set theory including the axiom of choice. We refer the reader to [De1], [Fe1], or [Je1] for any requisite background material. For the most part, our notation will be that commonly employed in set theory. An ordinal is, accordingly, identified with the set of all its predecessors, and a cardinal is defined to be an ordinal not equinumerous with any smaller ordinal. In general, α, β, γ, ... denote ordinals, with $\varkappa$ being reserved for cardinals. On denotes the class of all ordinals. If $\alpha = \beta+1$ for some β we say α is a <u>successor ordinal</u>, written $\mathrm{succ}(\alpha)$. Otherwise $\alpha = \bigcup\alpha$ and we say α is a <u>limit ordinal</u>, written $\lim(\alpha)$. A subset X of a limit ordinal α is said to be <u>cofinal</u> if $(\forall\xi\in\alpha)(\exists\zeta\in X)(\xi<\zeta)$.

If X is a set, $|X|$ denotes its cardinality. The <u>language of set theory</u>, LST, is the first order language (with equality) with the binary predicate $\in$ (set membership). We generally use φ, ψ to denote formulas in LST. If $F \subseteq A\times B$ and $X \subseteq B$, then $F''X$ denotes the set $\{x\in A \mid (\exists y\in X)[\langle x,y\rangle \in F]\}$. In particular, if $F: B \to A$ (so by definition $F \subseteq A\times B$) then $F''X = \{F(x) \mid x\in X\}$.

2. Constructibility

We shall assume the reader is familiar with the basic results concerning the construcible hierarchy, as described in, say, [De1]. In particular, we shall assume an acquaintance with the following material.

If X is a set, $\mathrm{Def}(X)$ denotes the set of all subsets of X which are first-order definable over X from parameters in X. Thus, $Z\in\mathrm{Def}(X)$ iff $Z\subseteq X$ and for some formula $\varphi(v_0,\ldots,v_n)$ of LST and

some $x_1,\dots,x_n \in X$, $Z = \{z \in X \mid \langle X,\in\rangle \models \varphi(z,x_1,\dots,x_n)\}$.
The <u>constructible hierarchy</u>, $\langle L_\alpha \mid \alpha \in On\rangle$, is defined by the recursion:

$$L_0 = \emptyset ;$$

$$L_{\alpha+1} = Def(L_\alpha) ;$$

$$L_\lambda = \cup_{\alpha<\lambda} L_\alpha \text{ , if } \lim(\lambda) .$$

The <u>constructible universe</u> is the class $L = \cup_{\alpha\in On} L_\alpha$. The collection of all sets, the <u>universe</u>, is denoted by V ; hence the formula $V = L$ (which can clearly be written in LST in the form $\forall x \exists \alpha (x \in L_\alpha)$ asserts that all sets are constructible. We refer to $V = L$ as the <u>axiom of constructibility</u>. There is a canonical L-definable well-ordering, $<_L$, of L with the property that for any limit ordinal α , $L_\alpha = <_L{}''L_\alpha$ and $<_L \cap L_\alpha^2$ is L_α-definable by the same formula which defines $<_L$ in L . For further details, we refer the reader to [De1]. In particular, he will there find a proof of the following results:

<u>Theorem 1</u>
If ZFC is consistent, so is ZFC + V = L . (Hence, any result which we prove using the assumption $V = L$ will be consistent relative to ZFC .)

<u>Theorem 2</u> (Condensation Lemma)
Let $\lim(\alpha)$. If $X \prec L_\alpha$ then $X \cong L_\beta$ for a unique $\beta \leq \alpha$.

<u>Theorem 3</u>
Let $\alpha_1 = \omega_1^L$, $\alpha_2 = \omega_2^L$. If $X \prec L_{\alpha_1}$, then $X = L_\beta$ for some $\beta \leq \alpha_1$.
If $X \prec L_{\alpha_2}$, then $X \cap L_{\alpha_1} = L_\beta$ for some $\beta \leq \alpha_1$.

If A is a given set we can define a hierarchy $\langle L_\alpha[A] \mid \alpha \in On\rangle$ of sets analogously to the definition of the constructible hierarchy, by allowing A to figure in the construction as a unary predicate in all of the definitions. (Thus $L_{\alpha+1}[A]$ will consist of all subsets of

$L_\alpha[A]$ which are first-order definable from members of $L_\alpha[A]$ in the structure $\langle L_\alpha[A], \in, A \cap L_\alpha[A]\rangle$. Full details of this construction are given in [De1], where proofs of the following analogues of theorems 2 and 3 may be found.

Theorem 4

Suppose $A \subseteq \omega_1^{L[A]}$. If $X \prec L_\alpha[A]$, then $X \cong L_\beta[A \cap \gamma]$ for unique ordinals $\gamma \leq \beta \leq \alpha$.

Theorem 5

Let $\alpha_1 = \omega_1^{L[A]}$, $\alpha_2 = \omega_2^{L[A]}$, $A \subseteq \alpha_1$. If $X \prec L_{\alpha_1}[A]$ then $X = L_\beta[A]$ for some $\beta \leq \alpha_1$. If $X \prec L_{\alpha_2}[A]$ then $X \cap L_{\alpha_1}[A] = L_\beta[A]$ for some $\beta \leq \alpha_1$.

3. Forcing

For the most part, it will be assumed that the reader is familiar with the method of forcing, as described in, say, chapter 17 of [Je1]. To fix the notation, however, we shall give here a very brief survey of this method.

Let $\mathbb{P} = \langle \mathbb{P}, \leq \rangle$ be a poset (partially ordered set). If $p,q \in \mathbb{P}$, we say p is an extension of q iff $p \leq q$. $\mathbb{P}$ is atomless if there is no $p \in \mathbb{P}$ which has no proper extension in $\mathbb{P}$. $K \subseteq \mathbb{P}$ is an initial section of $\mathbb{P}$ iff $(\forall p \in K)(\forall q \in \mathbb{P})(q \leq p \rightarrow q \in K)$; similarly final section. We say $D \subseteq \mathbb{P}$ is dense in $\mathbb{P}$ if every $p \in \mathbb{P}$ has an extension in D . If $p,q \in \mathbb{P}$, we say p and q are compatible iff p and q have a common extension; otherwise p and q are incompatible, written $p|q$. If $\varkappa$ is a cardinal, we say $\mathbb{P}$ has the $\varkappa$ chain condition if $\mathbb{P}$ has no pairwise incompatible subset of cardinality $\varkappa$. We refer to the ω_1 chain condition as the countable chain condition (c.c.c.). $\mathbb{P}$ is splitting if every $p \in \mathbb{P}$ has two incompatible extensions in $\mathbb{P}$.

Let P be a poset, X any set. A set $G \subseteq \mathbb{P}$ is said to be X-generic for $\mathbb{P}$ if G is a pairwise compatible, final section of $\mathbb{P}$ such that whenever $D \in X$ is a dense initial section of $\mathbb{P}$, then $D \cap G \neq \emptyset$. It is proved in [Je1] that if $|X \cap \mathcal{P}(\mathbb{P})| \leq \omega$, then there is a set G which is X-generic for $\mathbb{P}$.

Given a countable transitive model (c.t.m.) M of ZFC and a splitting poset $\mathbb{P} \in M$, if G is an M-generic set for $\mathbb{P}$, then $M[G]$, the constructible closure of $M \cup \{G\}$, is a c.t.m. of ZFC (called a generic extension of M) such that $On \cap M = On \cap M[G]$.

The method of forcing allows us to discuss the properties of $M[G]$ within M. (In particular, the proof that $M[G]$ is a model of ZFC uses the method of forcing.) We briefly outline this method below, leaving Jech [Je1] to supply all of the details. If $\mathbb{P}$ is a splitting poset, there is a complete boolean algebra associated with $\mathbb{P}$ in a natural way. Define a topology on $\mathbb{P}$ by taking as an open basis the collection of all subsets of $\mathbb{P}$ of the form $[p] = \{q \in \mathbb{P} \mid q \leq p\}$ for $p \in \mathbb{P}$. We denote by $BA(\mathbb{P})$ the (complete) boolean algebra of all regular open subsets of $\mathbb{P}$ under this topology. (Recall that $X \subseteq \mathbb{P}$ is regular open iff X is the interior of its closure.) The map $\pi : \mathbb{P} \to BA(\mathbb{P})$ defined by $\pi(p)$ = interior (closure ($[p]$)) is a one-one order-embedding of $\mathbb{P}$ into $BA(\mathbb{P})$, whose range is a dense subset of $BA(\mathbb{P})$ (when we regard $BA(\mathbb{P})$ as a poset in the usual way). We usually assume, therefore, that $BA(\mathbb{P})$ is isomorphed so that π is the identity map here, whence $\mathbb{P}$ is a dense subset of $BA(\mathbb{P})$ and the ordering of $\mathbb{P}$ is just the boolean ordering of $BA(\mathbb{P})$. If $\mathbb{P}$ is an element of a c.t.m. M of ZFC, then $\mathbb{B} = [BA(\mathbb{P})]^M$ is an M-complete boolean algebra which lies in M. The boolean universe $M^{(\mathbb{B})}$ is the set of all functions $u \in M$ such that $dom(u) \subseteq M^{(\mathbb{B})}$ and $ran(u) \subseteq \mathbb{B}$. ($M^{(\mathbb{B})}$ is defined by a simple recursion on rank within M.) $M^{(\mathbb{B})}$ is a definable class of M, and the (M-definable) map $\check{}$ defined recursively by $\check{a} = \{\langle 1, \check{x}\rangle \mid x \in a\}$ is a one-one embedding of M into

$M^{(\mathbb{B})}$. For each sentence φ of the language of set theory with parameters from $M^{(\mathbb{B})}$ we can define (in M) a unique boolean (truth-) value $\|\varphi\| \in \mathbb{B}$ in a natural way. If φ is a theorem of ZFC , then $\|\varphi\| = 1\!\!1$. The relationship $\|x=y\| = 1\!\!1$ defines an equivalence relation on $M^{(\mathbb{B})}$. If we wish, we can factor $M^{(\mathbb{B})}$ by this relation so that $\|x=y\| = 1\!\!1$ implies $x = y$. In this case we say $M^{(\mathbb{B})}$ is normalised.

Suppose now that G is M-generic for $\mathbb{P}$. Then G^{-} , the final section of $\mathbb{B}$ generated by G , will be an M-complete ultrafilter on $\mathbb{B}$ (and conversely). G^{-} determines a factorisation of $M^{(\mathbb{B})}$ into the c.t.m. $M[G]$ of ZFC in the obvious way (an atomic formula φ will become true in $M[G]$ just in case $\|\varphi\| \in G^{-}$). Each member of $M[G]$ thus corresponds to some G^{-}-equivalence class of members of $M^{(\mathbb{B})}$. We call any member of the equivalence class which determines $a \in M[G]$ a name for a . We usually use $\mathring{a}$ to denote an arbitrary name for a . Since $M[G]$ is just the constructible closure of $M \cup \{G\}$, it suffices to consider names for G and the elements of M only. The element $\mathring{G} = \{\langle b,\check{b}\rangle \mid b \in \mathbb{B}\}$ of $M^{(\mathbb{B})}$ is easily seen to be a name for G^{-} (which, it will be observed, is independent of the actual choice of G), and if $a \in M$, the element $\check{a}$ of $M^{(\mathbb{B})}$ is easily seen to be a name for a . The language determined by this particular collection of names is called the forcing language for $\mathbb{P}$. Since $\mathbb{P}$ is a dense subset of $\mathbb{B}$, it is easily seen that for any formula φ of the language of set theory, with parameters from $M[G]$, $M[G] \models \varphi$ iff $(\exists p \in G)(p \leq \|\mathring{\varphi}\|)$, where $\mathring{\varphi}$ is obtained from φ by replacing each parameter in φ by its name in the forcing language for $\mathbb{P}$. We define the forcing relation, $\Vdash$, between members of $\mathbb{P}$ and sentences $\mathring{\varphi}$ of the forcing language for $\mathbb{P}$ by $p \Vdash \mathring{\varphi}$ iff $p \leq \|\mathring{\varphi}\|$. In most "forcing arguments" we do not have to worry about the precise nature of $\mathbb{B}$ and the boolean values of various sentences;

rather a knowledge of $\mathbb{P}$ and of the forcing relation is adequate. That we can know $\Vdash$ without considering the boolean valuation of the forcing language follows from the easily established result that $p \Vdash \mathring{\varphi}$ iff $M[G] \models \varphi$ for all M-generic sets G (for $\mathbb{P}$) such that $p \in G$.

As a matter of notation, we write $\Vdash \mathring{\varphi}$ or $\mathbb{P} \Vdash \mathring{\varphi}$ iff $(\forall p \in \mathbb{P})(p \Vdash \mathring{\varphi})$. (Clearly, $\Vdash \mathring{\varphi}$ iff $\|\mathring{\varphi}\| = \mathbb{1}$.) Note that, by its definition, the relation $\Vdash$ is M-definable, a frequently used fact.

For the most part, we shall only be interested in <u>cardinal absolute</u> generic extensions. If M is a c.t.m. of ZFC, a generic extension $M[G]$ of M is said to be <u>cardinal absolute</u> if, for all ordinals α in M, $M \models$ "α is a cardinal" iff $M[G] \models$ "α is a cardinal". There are several criteria which give cardinal absolute extensions. The following are the ones we shall use. For proofs, see [Je1].

<u>Lemma 6</u>

If $\mathbb{P}$ is a (splitting) poset in a c.t.m. M of ZFC and $\kappa \in M$ is a regular uncountable cardinal in M, and if $M \models$ "$\mathbb{P}$ satisfies κ-c.c.", then for all $\alpha \in M$, such that $\alpha \geq \kappa$, $M \models$ "α is a cardinal" iff $M[G] \models$ "α is a cardinal", whenever G is M-generic for $\mathbb{P}$. In particular, if $\mathbb{P}$ satisfies c.c.c. in M, then $M[G]$ is a cardinal absolute extension of M.

Let κ be a cardinal, $\mathbb{P}$ a (splitting) poset. $\mathbb{P}$ is said to be <u>κ-closed</u> if, whenever $\langle p_\alpha \mid \alpha < \beta < \kappa \rangle$ is a decreasing sequence from $\mathbb{P}$, there is $p \in \mathbb{P}$ such that $\alpha < \beta \rightarrow p \leq p_\alpha$. $\mathbb{P}$ is <u>κ-dense</u> iff the intersection of any collection of fewer than κ dense initial sections of $\mathbb{P}$ is dense. Clearly, if $\mathbb{P}$ is κ-closed then $\mathbb{P}$ is κ-dense. For historical reasons we sometimes refer to "ω_1-closed" and "ω_1-dense" as "σ-closed" and "σ-dense", respectively.

Lemma 7

If $\mathbb{P}$ is a (splitting) poset in a c.t.m. M of ZFC and $\kappa \in M$ is a regular uncountable cardinal in M, and if $M \models$ "$\mathbb{P}$ is κ-dense", then for all $\alpha < \kappa$, $[M^\alpha]^M = [M^\alpha]^{M[G]}$, whenever G is M-generic for $\mathbb{P}$. In particular, for all $\alpha \leq \kappa$, $M \models$ "α is a cardinal" iff $M[G] \models$ "α is a cardinal", and moreover if $\alpha < \kappa$, then $\mathcal{P}^{M[G]}(\alpha) = \mathcal{P}^M(\alpha)$.

Since we shall only be concerned with splitting posets we shall henceforth supress the explicit mention of this adjective. We shall also make the (harmless, and usually true anyway) assumption that our posets have a unique maximum member, $\mathbb{1}$.
Note also that in the context of forcing, the elements of a poset $\mathbb{P}$ are often referred to as conditions.

Finally, we shall require the following result of Solovay. For a proof, see [Sh1].

Lemma 8 (Product Lemma)

Let $\mathbb{P}_1, \mathbb{P}_2$ be posets in the c.t.m. M of ZF. In M, define a poset $\mathbb{P}_1 \times \mathbb{P}_2$ on the cartesian product of $\mathbb{P}_1$ and $\mathbb{P}_2$ by $\langle p_1, p_2 \rangle \leq \langle q_1, q_2 \rangle \leftrightarrow p_1 \leq_1 q_1 \;\&\; p_2 \leq_2 q_2$. Then $G_1 \times G_2$ is M-generic for $\mathbb{P}_1 \times \mathbb{P}_2$ iff G_1 is M-generic for $\mathbb{P}_1$ and G_2 is $M[G_1]$-generic for $\mathbb{P}_2$, and in this case we have $M[G_1 \times G_2] = M[G_1][G_2]$. Furthermore, if G is M-generic for $\mathbb{P}_1 \times \mathbb{P}_2$, then $G_1 = \{p \in \mathbb{P}_1 \mid \langle p, \mathbb{1} \rangle \in G\}$ is M-generic for $\mathbb{P}_1$, $G_2 = \{q \in \mathbb{P}_2 \mid \langle \mathbb{1}, q \rangle \in G\}$ is $M[G_1]$-generic for $\mathbb{P}_2$, and $G = G_1 \times G_2$.

Chapter II

SOUSLIN'S HYPOTHESIS

1. Souslin lines

Let $S = \langle S, \leq \rangle$ be a linearly ordered set. We say S is dense, or a densely ordered set if whenever $x,y \in S$ and $x < y$ there is $z \in S$ such that $x < z < y$. Let S be an arbitrary densely ordered set for the rest of these definitions. S is complete if every subset A of S which has an upper bound in S has a least upper bound in S (denoted by $\mathrm{lub}(A)$). Note that there is an element of duality here. If S is complete, then every subset A of S having a lower bound in S has a greatest lower bound in S (denoted by $\mathrm{glb}(A)$), and conversely. Let us also remark that S is complete iff S is connected in the topology induced by the ordering. We say S is an ordered continuum if S is complete and has no end-points. $I \subseteq S$ is an interval if $I \neq \emptyset$ and $\forall x,y \in I \ \forall z \in S \ (x < z < y \rightarrow z \in I)$. An interval I is open if it has no end-points. I is closed if every point in $S - I$ is contained in an open interval disjoint from I. A subset D of S is dense in S if every open interval of S contains an element of D. (This is just the usual notion of dense subset under the topology on S, of course.) We say S is separable if it has a countable dense subset. The following elementary fact is well known.

Theorem 1

Every separable ordered continuum S is isomorphic to $\mathbb{R}$, the real line with the usual ordering.

One proves this theorem by showing that the countable dense subset D of S must be isomorphic to the set of rationals, $\mathbb{Q}$, and that the Dedekind completion of D is S itself. (The Dedekind completion of D is obtained by adding a lub to every cut in D without a lub. A cut is a proper initial segment of D.)

In 1920 M. Souslin [Su 1] raised the question as to whether one could obtain the same result with the separability requirement replaced by the following weaker assumption, which we will call the Souslin property:

> Every family of pairwise disjoint open intervals in S is countable.

"Souslin's hypothesis" is the assertion that this question has a positive answer, i.e.

Souslin's hypothesis (SH):

Every ordered continuum with the Souslin property is isomorphic to $\mathbb{R}$.

The major aim of this book is to show that Souslin's hypothesis is neither provable nor refutable in ZFC , even in the presence of GCH or its negation.

Theorem 2

The following are equivalent:

(i) Souslin's hypothesis,

(ii) every densely ordered set with the Souslin property can be (order-) embedded into $\mathbb{R}$,

(iii) every densely ordered set with the Souslin property has a countable dense subset.

Proof: (i) → (ii) and (iii) → (i) are trivial. For (ii) → (iii), let S be a densely ordered set, and let $f : S \to \mathbb{R}$ be an embedding given by (ii). Let $\langle q_i \mid i \in \omega \rangle$ be an enumeration of $\mathbb{Q}$. For each pair $i,j \in \omega$, choose $d_{ij} \in S$ such that $q_i < f(d_{ij}) < q_j$ if possible, otherwise d_{ij} is chosen arbitrarily. Then $D = \{d_{ij} \mid i,j \in \omega\}$ is a countable dense subset of S . ∎

Let us call an ordered continuum a Souslin line if it has the Souslin property but is not separable. Thus there exists a Souslin line if and only if Souslin's hypothesis is false.

Let S be a Souslin line. We state some simple properties of S, which tell us that SH is, to some extent, a natural conjecture. In Theorem 4 below we will show that S has a dense subset D of cardinality ω_1. If now $\langle x_\nu \mid \nu < \eta \rangle$ is a strictly increasing sequence in S, then $\eta < \omega_1$ (for otherwise $\{\langle x_\nu, x_{\nu+1}\rangle \mid \nu < \omega_1\}$ is an uncontable family of pairwise disjoint open intervals). Consequently:

(1) For each point $x \in S$ there exists a strictly increasing sequence $\langle x_i \mid i \in \omega \rangle$ from D with x as the least upper bound.

(2) S has cardinality 2^ω. (Since S is complete, $|S| \geq 2^\omega$. Conversely, by (1), there exists an injection from S into $\mathcal{P}_{\omega_1}(D)$, the set of all countable subsets of D, a set of cardinality 2^ω.)

(3) For each $x \in S$ there exists a strictly increasing sequence $\langle x_i \mid i \in \omega \rangle$ from S with lub x. For $n \in \omega$, let $f_n : S \to S$ be defined by $f_n(x) = x_n$. Then the functions f_n are not all continuous, by a theorem of J. Ball [Ba 1].

In conclusion we mention one topological question which historically has been connected with SH. Mary Ellen Rudin [Ru 1] has shown that if there exists a Souslin line, then there exists a normal Hausdorff space which is not countably paracompact. Together with the result of C.H. Dowker [Do 1] that a topological space X is countably paracompact and normal iff $X \times I$ is normal (I is the real unit interval), we obtain that in ZFC + $\neg$SH the following is provable:

(4) There exists a normal Hausdorff space X such that $X \times I$ is not normal.

Recently, M.E. Rudin [Ru 3] has found a proof of (4) without assuming $\neg SH$, i.e. as a theorem in ZFC alone.

2. Souslin trees

A tree is a poset $\underset{\sim}{T} = \langle T, \leq \rangle$ such that for every $x \in T$, the set $\hat{x} = \{y \in T \mid y < x\}$ is well ordered by $\leq$. Let $\underset{\sim}{T}$ be a tree. For $x \in T$, the height of x , $ht(x)$, is the order type of $\hat{x}$ under $\leq$. (Hence we may define $ht(x)$ by induction as follows: $ht(x) = \sup\{ht(y) \mid y<x\}$.) For $\alpha \in On$, the α'th level of $\underset{\sim}{T}$ is the set $T_\alpha = \{x \in T \mid ht(x) = \alpha\}$. We write $\underset{\sim}{T}|\alpha$ for the restriction of $\underset{\sim}{T}$ to the set $T|\alpha = \bigcup_{\beta<\alpha} T_\beta$. For $x \in T$ we set $T^x = \{y \in T \mid x \leq y\}$. When $X \subseteq T$ we let $ht(X) = \sup\{ht(x) \mid x \in X\}$. $ht(T)$ is called the length of $\underset{\sim}{T}$. (From now on, when there is no danger of confusion, we denote a tree simply by $T = \langle T, \leq \rangle$.)

Two elements x,y of a tree T are comparable if $x \leq y$ or $y \leq x$; if not, they are incomparable, written $x|y$. A chain is a linearly ordered subset of T . A branch b is a chain which is $\leq$-closed, i.e. if $x \in b$ and $y \leq x$, then $y \in b$. If $ht(b) = \alpha$, then b is an α-branch. An antichain is a set of pairwise incomparable elements of T . A branch (resp. an antichain) is called maximal if it is not included in a larger branch (resp. antichain).

A tree T is called an Aronszajn tree if $|T| = \omega_1$ and every chain in T and every level T_α is countable . T is called a Souslin tree if $|T| = \omega_1$ and every chain and antichain in T is countable.

Note that if T is Aronszajn or Souslin, then $ht(T) = \omega_1$. Since each T_α is an antichain, every Souslin tree is Aronszajn. Aronszajn trees can (in the presence of the axiom of choice) be proved to exist (c.f. [Je 1], [Je 2], or [Ru 2]). It is often convenient that the trees satisfy certain "normality" properties. So let us call a tree T a normal tree of length α (or a normal α-tree) if:

(i) $ht(T) = \alpha$,

(ii) T has a unique least point, called the root of T .

(iii) each non-maximal point x has at least two immediate successors (i.e. points y such that $ht(y) = ht(x) + 1$ and $x \leq y$),

(iv) each point has successors at each greater level $< \alpha$,

(v) each β-branch has at most one immediate successor when β is a limit ordinal,

(vi) each level T_ν is countable.

If T is normal, then $ht(T) \leq \omega_1$ (otherwise, by (iii) and (iv), T_{ω_1} will be uncountable). A normal Aronszajn tree is a normal ω_1-tree without uncountable chains. A normal Souslin tree is a normal ω_1-tree without uncountable antichains. Then a normal Souslin tree is indeed a Souslin tree. (For an uncountable chain gives rise to an uncountable antichain, again by (iii).)

Theorem 3

Every Souslin tree T can be normalized, i.e. T contains a normal Souslin tree T^* . Similarly for Aronszajn trees.

Proof: Remove from T the points x for which T^x is countable, and then remove those points with only one immediate successor. The resulting tree, T' , is still of length ω_1 , hence Souslin. By induction, pick out one point in T_0 and one point extending every branch of limit length with extensions in T' . The resulting subtree, T^* , is a normal Souslin tree. The same construction applies also for Aronszajn trees. ∎

The long list of definitions above is justified by the following fact, due to E.W. Miller [Mi 1] :

Theorem 4

Souslin's hypothesis is equivalent to the non-existence of a Souslin tree, i.e. there exists a Souslin line if and only if there exists a Souslin tree. Furthermore, every Souslin line has a dense subset of cardinality ω_1 .

Proof: Suppose S is a Souslin line. Observe that if a subset D of S is countable, it is not dense in S, whence $S - \bar{D}$ is the union of a non-empty collection $I(D)$ of pairwise disjoint open intervals. And of course, $I(D)$ is countable, since S has the Souslin property. ($\bar{D}$ above denotes the topological closure of D in S.) Let D_o consist of one point from S. By induction, assume D_β is defined for $\beta < \alpha$. Set $I_\alpha = I(\bigcup_{\beta<\alpha} D_\beta)$ and let D_α consist of one point from each interval in I_α . Now let $T = \bigcup_{\alpha<\omega_1} I_\alpha$ and partially order T by reverse inclusion. We claim that $\langle T, \supseteq \rangle$ is a Souslin tree.

Indeed, T is a tree of cardinality ω_1 with "normality property" (iii) satisfied, hence it is enough to show that every antichain in T is countable. But an antichain in T is just a family of pairwise disjoint open intervals in S , hence countable.

Now we look at $D = \bigcup_{\alpha<\omega_1} D_\alpha$. Clearly, $|D| = \omega_1$, and we claim that D is dense in S . If not, let $p \in S - \bar{D}$. For each $\alpha < \omega_1$ there is an interval $J_\alpha \in I_\alpha$ containing p . But $J_\alpha \supset J_{\alpha+1}$ for each α , so $\{J_\alpha \mid \alpha < \omega_1\}$ is an uncountable chain in T , impossible! Hence we have shown that a Souslin line has a dense subset of cardinality ω_1 .

On the other hand, suppose that $\langle T, \leq_T \rangle$ is a normal Souslin tree. In fact, we may also assume that T satisfies the following strengthening of (iii): Every point in T has ω many immediate successors. (Such a tree may be obtained from a normal Souslin tree T' by taking the restriction of T' to the points at limit levels in T'.) For $\alpha > 0$, let $\leq_\alpha$ be a linear ordering of T_α of order type that of the rationals, such that if $\alpha < \beta$, $x,y \in T_\alpha$, $x',y' \in T_\beta$, $x < x'$, $y < y'$ and $x <_\alpha y$, then $x' <_\beta y'$. Let S be the set of maximal branches of T. When b is a branch of T, we let b_α be the point in b which has height α. We order S by $\leq$ as follows: $b < d$ iff $b_\alpha <_\alpha d_\alpha$, when $\alpha = \alpha(b,d)$ is the least ordinal such that $b_\beta = d_\beta$ for all $\beta < \alpha$ and $b_\alpha \neq d_\alpha$. Then $\langle S, \leq \rangle$ is a densely ordered set. In view of Theorem 2, it suffices to show that S has the Souslin property and no countable dense subset.

For $x \in T$, let $I_x = \{b \in S \mid x \in b\}$. To every open interval I there is an $x = x_I \in T$ such that $I_x \subseteq I$. (For if $I = \langle b,d \rangle = \{c \in S \mid b < c < d\}$, let $\alpha = \alpha(b,d)$. Then we may take $x \in T_\alpha$ such that $b_\alpha <_\alpha x <_\alpha d_\alpha$.) When I and J are disjoint, then x_I and x_J are incomparable. Hence an uncountable family of pairwise disjoint open intervals in S would give rise to an uncountable antichain in T!

Finally, suppose that D is a countable dense subset of S. Let $\alpha = \sup\{ht(b) \mid b \in D\}$. Then $\alpha < \omega_1$, so if we take $x \in T_\alpha$, then I_x contains an open interval disjoint from D. Contradiction! ▌

By this theorem, we may concentrate our interest on the Souslin trees. We shall, however, return to the Souslin line in Chapters IV and V, where we transform certain homogenity properties of Souslin trees to corresponding properties of Souslin lines.

For later use, we note the following definitions. A normal tree T of length ω_1 is called a special Aronszajn tree if T is the union of a countable number of its antichains. (Any such tree is clearly Aronszajn.) When $T = \langle T, \leq_T \rangle$ is a tree and $X = \langle X, \leq_X \rangle$ is a poset, we say that T is X-embeddable if there is $f : T \to X$ such that $x <_T y \to f(x) <_X f(y)$. We then say f embeds T in X.

Theorem 5

Let T be a normal ω_1-tree.

(i) T is $\mathbb{Q}$-embeddable iff T is a special Aronszajn tree.

(ii) If T is $\mathbb{R}$-embeddable, then T is Aronszajn, and every uncountable subset of T contains an uncountable antichain of T.

Proof: (i) Let f embed T in $\mathbb{Q}$. For each $q \in \mathbb{Q}$, let $A_q = \{x \in T \mid f(x) = q\}$. Then every A_q is an antichain in T and $T = \bigcup_{q\in\mathbb{Q}} A_q$. Conversely, let $T = \bigcup_{n\in\omega} A_n$ where the antichains are pairwise disjoint.

Define an embedding $f : T \to \mathbb{Q}$ as follows: For each $x \in A_0$, let e.g. $f(x) = 0$. For $x \in A_1$, let $f(x) = 1$ (resp. $f(x) = -1$) if x does not lie beneath (resp. lies beneath) a point of A_0. Go on like this, setting for instance $f(x) = -\frac{1}{2}$ if x is a point of A_2 lying above A_1 but beneath A_0. You'll never fail!

(ii) If T is $\mathbb{R}$-embeddable, then every branch of T must be countable, hence T is Aronszajn. Let $U \subseteq T$ be uncountable. Then U inherits a tree structure from T, with new levels U_α. We may assume that $U_\alpha \neq \emptyset$ for all $\alpha < \omega_1$, for otherwise we are already done, since then one of the countably many non-empty U_α's must be uncountable. Let $U^* = \bigcup_{\alpha<\omega_1} U_{\alpha+1}$. If now f embeds T (and hence also U) in $\mathbb{R}$, let g be such that $g(x) \in \mathbb{Q}$ and $f(y) < g(x) \leq f(x)$ whenever x is in some $U_{\alpha+1}$ and y is its predecessor in U_α. Then g embeds U^* in $\mathbb{Q}$. Hence U^* is the union of countably many antichains, one of which must be uncountable. ∎

By this theorem, no Souslin tree can be $\mathbb{R}$-embeddable. Hence we may ask: Can there be Aronszajn trees $\mathbb{R}$-embeddable but not $\mathbb{Q}$-embeddable? In fact, the Aronszajn trees constructed in ZFC are all special Aronszajn. But J. Baumgartner has proved [Bu 1] that in L there is an $\mathbb{R}$-embeddable non-special Aronszajn tree. Later in this book we will show that it is consistent with ZFC to assume that every Aronszajn tree is special Aronszajn (thereby obtaining SH).

3. Souslin trees in L .

Hitherto, we have not said anything definite concerning the truth or falsity of Souslin's hypothesis. We now turn towards consistency proofs. It was Tennenbaum who first proved the consistency of ¬SH . Later, Jech independently gave another proof. Their result is that any c.t.m. of ZFC may be extended by forcing to a model of ¬SH . In Tennenbaum's proof, CH holds in the extension if and only if it holds in the ground model. Hence ZF + GCH + ¬SH and ZFC + ¬CH + ¬SH are consistent relative to ZF . For details of these results, we refer the reader to Jech [Je 2].

We shall here obtain these consistency results in another way. In this section we prove the well-known result of R.B. Jensen that L is a model of ¬SH . In the next section we use this result to obtain the consistency of ¬CH + ¬SH relative to ZFC .

Theorem 6

Assume $V = L$. Then there exists a Souslin tree. Hence SH fails in L .

Proof: We are going to define a normal tree T of length ω_1 such that $T_\alpha \subseteq 2^\alpha$ for $\alpha < \omega_1$. The ordering will be ordinary inclusion (i.e. the initial sequence ordering). By induction on α we define the levels as follows:

T_0 consists of the empty sequence $\emptyset$ alone.

$T_{\alpha+1}$ consists of the sequences $s^\frown\langle i\rangle$ ($^\frown$ means concatenation of sequences) for $s \in T_\alpha$ and $i \in 2$.

Now, assume α is a limit ordinal $< \omega_1$. Let δ_α be the least δ such that

$$T|\alpha \in L_\delta$$
$$L_\delta \models ZF^-, \text{ and}$$
$$L_\delta \models \alpha \text{ is countable.}$$

(ZF^- means ZF minus the power set axiom.) Then $\delta_\alpha < \omega_1$, so L_{δ_α} contains only countably many maximal antichains A_n, $n \in \omega$, of $T|\alpha = \bigcup_{\beta<\alpha} T_\beta$. Hence, for every $s \in T|\alpha$ there exists an α-branch b going through s and meeting every A_n. (Such a branch may be obtained as follows. Let $\langle \alpha_n \mid n \in \omega\rangle$ be a strictly increasing sequence such that $\sup_{n\in\omega} \alpha_n = \alpha$. Define by induction a sequence $\langle s_n \mid n \in \omega\rangle$ such that $s_0 \geq_T s$, and for each n, s_n lies above a point in A_n, $ht(s_n) \geq \alpha_n$, and $s_{n+1} \geq_T s_n$. Then let $b = \{t \mid \exists n (t \leq_T s_n)\}$.)

Let b_s be the least such branch in the canonical well-ordering of L. Let T_α consist of the sequences $\bigcup b_s$ for $s \in T|\alpha$. Hence T is defined.

T is obviously a normal tree of length ω_1, so it remains to show that every antichain of T is countable. Let A be a maximal antichain of T. Now $T, A \subset L_{\omega_1}$, so $T, A \in L_{\omega_2}$.

Let M be a countable, elementary submodel of $\langle L_{\omega_2}, \in\rangle$ such that $A \in M$. We also want $T, \omega_1 \in M$, but this is automatically satisfied since both are definable in L_{ω_2} without parameters. Now we use Theorem I.2. It follows that $\omega_1 \cap M$ is transitive, so let $\alpha = \omega_1 \cap M < \omega_1$. Let π, β be such that

$$\pi : M \cong \langle L_\beta, \in \rangle .$$

π , the collapsing isomorphism, is defined by $\pi(x) = \pi''(x \cap M)$ for $x \in M$. Hence if $x \in L_{\omega_1} \cap M$, then $x \subseteq M$, so $\pi(x) = x$ (by induction). Thus $\pi(\omega_1) = \alpha$. Now, T_γ is definable in L_{ω_1} in the parameter γ for $\gamma < \omega_1$. Thus $T_\gamma \in M$ for $\gamma < \alpha$. So $\pi(T_\gamma) = T_\gamma$. Hence we obtain

$$\pi(T) = \pi(\bigcup_{\gamma<\omega_1} T_\gamma)$$

$$= \bigcup_{\gamma<\pi(\omega_1)} \pi(T_\gamma) = T|\alpha ,$$

since T is defined as the union of the T_γ's . Since $A \subset T$, $\pi(A) \subset T|\alpha$, so

$$\pi(A) = \pi''(A \cap M) = A \cap M = A \cap T|\alpha .$$

Now we are ready to conclude the proof. Since ω_1 is uncountable in L_{ω_2} and in M, $\alpha = \pi(\omega_1)$ is uncountable in L_β . But α is countable in L_{δ_α} . Hence $\beta < \delta_\alpha$. Also, since A is a maximal antichain of T, $A \cap T|\alpha$ is a maximal antichain of $T|\alpha$. But since $A \cap T|\alpha \in L_\beta \subseteq L_{\delta_\alpha}$, by the construction of T_α, every point in T_α lies above a point of $A \cap T|\alpha$. Thus $A \cap T|\alpha$ is maximal also in T. Hence $A = A \cap T|\alpha$, so A is countable. ▮

Note: By modifying the proof we may prove that if $V = L[A]$ for some $A \subseteq \omega_1$, then $\neg SH$ (c.f. [Je 2]).

<u>Corollary</u>: If ZF is consistent, then so is ZF + GCH + $\neg$SH .

A final remark. There are two other "axioms" which also decide the existence of Souslin trees. Martin's axiom (+ $\neg$CH) implies that there are no Souslin trees, hence SH . (This will be treated in Chapter VI.) The axiom of determinateness, AD , implies that ω_1 is measurable, and thus that there are no Souslin trees, But we cannot say that AD implies SH, since the proof of Theorem 4 uses heavily the axiom of choice.

4. Forcing and Souslin trees

Let M be a c.t.m. of ZFC, and assume $T = \langle T, \leq_T \rangle$ is a Souslin tree in M. We are interested in what will happen to T under forcing extensions. First of all, let us see the result when we force with T itself with reversed ordering, i.e. if we force with the poset $\langle \mathbb{P}, \leq \rangle = \langle T, \geq_T \rangle$. In this case the notions "compatible" and "comparable" coincides.

Thus a set of pairwise incompatible elements of $\mathbb{P}$ is just an antichain in T. Since all these are countable, $\mathbb{P}$ satisfies the c.c.c.. Hence generic extensions are cardinal absolute.

Now let G be M-generic for $\mathbb{P}$. Since the elements of G are compatible, G is a chain in T. G is a final section of $\mathbb{P}$, hence initial in T, so G is a branch. For each $\alpha < \omega_1^M$, $D_\alpha = \bigcup_{\beta \geq \alpha} T_\beta$ is clearly a dense initial section of $\mathbb{P}$ lying in M. Hence $G \cap D_\alpha \neq \emptyset$ for $\alpha < \omega_1^M$, so G is an ω_1^M-branch of T. For convenience, we call an ω_1^M-branch b <u>M-generic for T</u> if b is M-generic for $\mathbb{P}$. (In Chapter VI it will be convenient to change this notation slightly.) In the real world, T has countable length, namely ω_1^M, and so T has 2^ω ω_1^M-branches. We now prove that they all are M-generic for T.

Theorem 7

Let M be a c.t.m. of ZFC, and let T be a Souslin tree in M. Let $b \subseteq T$. Then b will be a cofinal branch of T just in case b is M-generic for T.

Proof: One half of the theorem is already proved. So suppose that b is a cofinal branch of T. Then b is a pairwise compatible final section of $\mathbb{P}$, so it clearly suffices to show that if D is a dense initial section of $\mathbb{P}$, then $D \supseteq D_\alpha$ for some

$\alpha < \omega_1^M$. (Notations as above.) Setting

$$A = \{x \in D \mid \neg \exists y \in D \; y <_T x\},$$

$A \in M$ and A is a maximal antichain in T, hence countable in M. So letting $\alpha = ht(A) = \sup\{ht(x) \mid x \in A\}$, we must have $D \supseteq D_\alpha$.

Thus we have found a method of "killing" a Souslin tree, since in the extension the tree contains an ω_1-branch, and is not even Aronszajn (but is still a normal tree of length ω_1). This killing procedure will be explored further in Chapter VI.

In fact, the construction of limit levels T_α in the proof of Theorem 6 is a "forcing" construction. An L_{δ_α}-generic subset of $\langle T|\alpha, \supseteq\rangle$ is an α-branch. The dense initial sections of $\langle T|\alpha, \supseteq\rangle$ lying in L_{δ_α} are precisely the sets $A_n^* = \{s \in T|\alpha \mid \exists t \in A_n \; s \supseteq t\}$, hence an α-branch is L_{δ_α}-generic iff it meets every A_n. Hence we could have defined the branch b_s as the $<_L$-least L_{δ_α}-generic branch going through s, for $s \in T|\alpha$.

In certain forcing constructions, all Souslin trees are preserved. (See for instance Lemma VII.4.) We now prove that if we destroy CH in the usual way, all Souslin trees are preserved.

Theorem 8

Let M be a c.t.m. of ZFC, and assume $T = \langle T, \leq_T\rangle$ is a Souslin tree in M. Let $\kappa > (2^\omega)^M$ be regular. Let $\mathbb{P}$ be the Cohen conditions to make $2^\omega = \kappa$ in the extension (i.e. $\mathbb{P}$ is the set of all finite maps from a subset of $\kappa \times \omega$ to 2, ordered by reverse inclusion). Let G be M-generic for $\mathbb{P}$. Then T is Souslin also in $M[G]$.

Proof: We first note that $\mathbb{P}$ satisfies c.c.c. in M, hence cardinals are absolute in the extension. By a standard "pigeon-hole" proof it is also easily shown that every uncountable subset of

$\mathbb{P}$ contains an uncountable set consisting of pairwise compatible elements.

Without loss of generality we may assume that T is a normal Souslin tree in M. (For if the theorem is true for the "normalized" tree, then it is easily seen to be true for the original tree.) Let $A \in M[G]$ be an antichain in T. It suffices to prove that A is countable in $M[G]$. Let $\mathring{A}$ be a name for A, and let $p^* \in G$ such that $p^* \Vdash$ "$\mathring{A}$ is an antichain of $\check{T}$". Let

$$B = \{t \in T \mid \exists p \leq p^* \ (p \Vdash \check{t} \in \mathring{A})\} .$$

Then $B \in M$ and $A \subseteq B$, so it suffices to show that B is countable in M.

From now on we work in M. For each $t \in B$, pick $p_t \leq p^*$ such that $p_t \Vdash \check{t} \in \mathring{A}$. Note that if $t \neq t'$ and p_t and $p_{t'}$ are compatible, then $t \mid t'$. (For if $p \leq p_t, p_{t'}$, then $p \Vdash$ "$\check{t} \neq \check{t}'$, $\check{t}, \check{t}' \in \mathring{A}$, and $\mathring{A}$ is an antichain", so $p \Vdash \check{t} \mid \check{t}'$.)

Suppose that B is uncountable. Let $C = \{p_t \mid t \in B\}$. If C is countable, then for some p_t the set $\{t' \mid p_t \Vdash \check{t}' \in \mathring{A}\}$ is an uncountable antichain of T, which is impossible! So C contains an uncountable subset D of pairwise compatible conditions. But then the set $\{t \mid p_t \in D\}$ is an uncountable antichain of T, and a contradiction is obtained. ▮

<u>Corollary</u>

If ZF is consistent, then so is ZFC + ¬CH + ¬SH .

Proof: $\bigcup G$ is a function from $\varkappa \times \omega$ to 2, and gives rise to $\varkappa > \omega_1^{M[G]}$ different subsets of ω. Hence $M[G] \models$ ZFC + ¬CH + ¬SH . ▮

In fact, an alternative proof of the above theorem may be based on the following general result (implicit in Lemma VI.3, and not so difficult to prove directly):

> Let M be a c.t.m. of ZFC, and let $\mathbb{P}, \mathbb{Q}$ be sets of conditions in M. Then the following are equivalent:
>
> (i) $M \models \mathbb{P} \times \mathbb{Q}$ sat. c.c.c.,
>
> (ii) $M \models \mathbb{P}$ sat. c.c.c., and $\mathbb{P} \Vdash \check{\mathbb{Q}}$ sat. c.c.c. (i.e. $\mathbb{Q}$ satisfies c.c.c. in $M[G]$ whenever G is M-generic for $\mathbb{P}$).

From this fact, Theorem 8 is an immediate consequence: Let $\mathbb{P}, T$ be as in the theorem. Since T is Souslin in M, and $\mathbb{P}$ satisfies c.c.c. in every model, $M \models T$ sat. c.c.c., and $T \Vdash \check{\mathbb{P}}$ sat. c.c.c.. Hence $M \models \mathbb{P} \times T$ sat. c.c.c., so $\mathbb{P} \Vdash \check{T}$ sat. c.c.c., i.e. $\mathbb{P} \Vdash \check{T}$ is Souslin.

Chapter III

THE COMBINATORIAL PROPERTY $\diamond$

1. Statement of $\diamond$

By analyzing the proof of Theorem II. 6 (that Souslin trees exist in L) it is possible to isolate as a principle the "combinatorial core" that makes the whole thing work. The definition of this principle, $\diamond$, is due to Jensen.

First some necessary definitions. Let λ be a limit ordinal. If $C \subseteq \lambda$, we say that C is closed (in λ) if $\bigcup(\alpha \cap C) \in C$ for all $\alpha < \lambda$. C is unbounded or cofinal in λ if $\forall \alpha < \lambda \ \exists \beta \in C \ (\alpha < \beta)$. $A \subseteq \lambda$ is stationary (or Mahlo) in λ if it intersects every closed unbounded subset of λ.

Assume now that $\varkappa > \omega$ is regular. It is easy to show that the intersection of less that $\varkappa$ closed unbounded subsets of $\varkappa$ is itself closed unbounded. As a first consequence it is clear that every closed unbounded subset of $\varkappa$ must be stationary. As a second consequence we obtain the following simple lemma:

Lemma 1

Let $A_n \subseteq \omega_1$ for $n \in \omega$. If $\bigcup_{n\in\omega} A_n$ is stationary in ω_1, then A_n is stationary in ω_1 for some $n \in \omega$.

Proof: If every A_n is not stationary, let C_n be closed unbounded in ω_1 such that $A_n \cap C_n = \emptyset$ for $n \in \omega$. Then $(\bigcup_{n\in\omega} A_n) \cap (\bigcap_{n\in\omega} C_n) = \emptyset$, so $\bigcup_{n\in\omega} A_n$ is not stationary in ω_1. ∎

By $\diamond$ we mean the following principle:

There is a sequence $\langle S_\alpha \mid \alpha < \omega_1 \rangle$ such that $S_\alpha \subseteq \mathcal{P}(\alpha)$ and

$|S_\alpha| \leq \omega$ for $\alpha < \omega_1$, and: If $X \subseteq \omega_1$, then the set $\{\alpha < \omega_1 \mid X \cap \alpha \in S_\alpha\}$ is stationary in ω_1.

Later, we shall see how (at the cost of some minor combinatorial details) we can split the proof of Theorem II. 6 into two parts, namely Theorem 4 and 5 below. The role of the L_{δ_α}'s will be taken over by the S_α's.

Theorem 2

$\diamondsuit$ is equivalent to the following property:

There is a sequence $\langle S_\alpha \mid \alpha < \omega_1 \rangle$ such that $S_\alpha \subseteq \alpha$ for $\alpha < \omega_1$ and if $X \subseteq \omega_1$, then the set $\{\alpha < \omega_1 \mid X \cap \alpha = S_\alpha\}$ is stationary in ω_1.

Proof: This proof is due to K. Kunen. Assume $\diamondsuit$.

Claim 1: There is a sequence $\langle S_\alpha^n \mid \alpha < \omega_1 \wedge n \in \omega \rangle$ such that each $S_\alpha^n \subseteq \alpha \times \omega$ and: If $X \subseteq \omega_1 \times \omega$, then $\{\alpha \mid \exists n\ X \cap (\alpha \times \omega) = S_\alpha^n\}$ is stationary.

Proof: Let the sequence $\langle \bar{S}_\alpha \mid \alpha < \omega_1 \rangle$ satisfy the clauses in $\diamondsuit$. Let $f : \omega_1 \to \omega_1 \times \omega$ be the order isomorphism between ω_1 and $\omega_1 \times \omega$ (ordered lexicographically). Let $f_* : \mathcal{P}(\omega_1) \to \mathcal{P}(\omega_1 \times \omega)$ be defined by $f_*(X) = f''(X)$. Let $S_\alpha =_{df} f_*''\bar{S}_\alpha$. Then $S_\alpha \subseteq \mathcal{P}(\alpha \times \omega)$. Furthermore, $|S_\alpha| \leq \omega$, so let $\langle S_\alpha^n \mid n \in \omega \rangle$ enumerate S_α for $\alpha < \omega_1$.

Now, let $X \subseteq \omega_1 \times \omega$. Set $A = \{\alpha \mid X \cap (\alpha \times \omega) \in S_\alpha\}$. Let C be closed unbounded in ω_1. We prove that $A \cap C \neq \emptyset$. Let $D = \{\omega^{\omega+\nu} \mid \nu < \omega_1\}$. Then D is closed unbounded, and we note that if $\alpha \in D$, then $f(\alpha) = \langle \alpha, 0 \rangle$, so $f_*(\alpha) = \alpha \times \omega$. Define $\bar{X} = f_*^{-1}(X)$ and $B = \{\alpha \mid \bar{X} \cap \alpha \in \bar{S}_\alpha\}$. By $\diamondsuit$, B is stationary, so let $\alpha \in B \cap C \cap D$. Then $f_*(\bar{X} \cap \alpha) = f_*(\bar{X}) \cap f_*(\alpha) = X \cap (\alpha \times \omega)$.

Since $\bar{X} \cap \alpha \in \bar{S}_\alpha$, $f_*(\bar{X} \cap \alpha) \in S_\alpha$, i.e. $X \cap (\alpha \times \omega) \in S_\alpha$. Hence $\alpha \in A \cap C$, and Claim 1 is proved.

Claim 2: There is an $n_o \in \omega$ such that for all $Y \subseteq \omega_1$ there exists $X \subseteq \omega_1 \times \omega$ such that $Y = X''\{n_o\}$ and $\{\alpha \mid X \cap (\alpha \times \omega) = S_\alpha^{n_o}\}$ is stationary.

Proof: Suppose the claim is false, and let, for $n \in \omega$, $Y_n \subseteq \omega_1$ be a counterexample to the claim. Let $X = \{\langle y,n \rangle \mid y \in Y_n \wedge n \in \omega\}$. Then $X''\{n\} = Y_n$ for all $n \in \omega$. Hence for all $n \in \omega$, $A_n = \{\alpha \mid X \cap (\alpha \times \omega) = S_\alpha^n\}$ is not stationary, so by Lemma 1, $\bigcup_{n \in \omega} A_n = \{\alpha \mid \exists n \in \omega \;\; X \cap (\alpha \times \omega) = S_\alpha^n\} = \{\alpha \mid X \cap (\alpha \times \omega) \in S_\alpha\}$ is not stationary, contradicting Claim 1!

Now, set $\hat{S}_\alpha = S_\alpha^{n_o}{}''\{n_o\}$ for $\alpha < \omega_1$. Then the sequence $\langle \hat{S}_\alpha \mid \alpha < \omega_1 \rangle$ satisfies the clauses in the theorem. Indeed, let $Y \subseteq \omega_1$ and let $C \subseteq \omega_1$ be closed unbounded. By Claim 2, let $X \subseteq \omega_1 \times \omega$ be such that $Y = X''\{n_o\}$ and $\{\alpha \mid X \cap (\alpha \times \omega) = S_\alpha^{n_o}\}$ is stationary. Pick $\alpha \in C$ such that $X \cap (\alpha \times \omega) = S_\alpha^{n_o}$. Then $Y \cap \alpha = X''\{n_o\} \cap \alpha = (X \cap (\alpha \times \omega))''\{n_o\} = S_\alpha^{n_o}{}''\{n_o\} = \hat{S}_\alpha$. ∎

Hereafter, by $\diamondsuit$ we mean the equivalent property stated in this theorem. We leave it to the reader to prove that $\diamondsuit$ is also equivalent to the following principle: There is a sequence $\langle h_\alpha \mid \alpha < \omega_1 \rangle$ of functions such that for every function $h : \omega_1 \to \omega_1$, the set $\{\alpha < \omega_1 \mid h \restriction \alpha = h_\alpha\}$ is stationary. For completeness and for later use we now state two stronger versions of $\diamondsuit$.

$\diamondsuit^*$: There is a sequence $\langle S_\alpha \mid \alpha < \omega_1 \rangle$ such that $S_\alpha \subseteq \mathcal{P}(\alpha)$ and $|S_\alpha| \leq \omega$ for $\alpha < \omega_1$, and: If $X \subseteq \omega_1$, then the set $\{\alpha \mid X \cap \alpha \in S_\alpha\}$ contains a closed unbounded set.

$\diamondsuit^+$: There is a sequence $\langle S_\alpha \mid \alpha < \omega_1 \rangle$ such that $S_\alpha \subseteq \mathcal{P}(\alpha)$ and

$|S_\alpha| \leq \omega$ for $\alpha < \omega_1$, and: For all $X \subseteq \omega_1$ there exists a closed unbounded $C \subseteq \omega_1$ such that if $\alpha \in C$, then $X \cap \alpha$, $C \cap \alpha \in S_\alpha$.

Clearly $\diamondsuit^+ \to \diamondsuit^* \to \diamondsuit$. Jensen has proved that the arrows cannot be reversed.

2. Properties of $\diamondsuit$

Theorem 3

$\diamondsuit$ implies CH.

Proof: Let $\langle S_\alpha \mid \alpha < \omega_1 \rangle$ satisfy the clauses of $\diamondsuit$. To every $a \subseteq \omega$ $(\subseteq \omega_1)$ there exists an $\alpha < \omega_1$ such that $a = S_\alpha$. Let $f(a)$ be the least such α. Then f is an injection from $\mathcal{P}(\omega)$ into ω_1. ∎

That $\diamondsuit$ is actually stronger than CH (which it strongly seems to be) will be (implicitly) shown in the last part of this book.

Theorem 4

Assume $V = L$. Then $\diamondsuit$ is valid.

Proof: By induction on $\alpha < \omega_1$ we define a sequence $\langle \langle S_\alpha, C_\alpha \rangle \mid \alpha < \omega_1 \rangle$ as follows. Set $S_0 = C_0 = \emptyset$ and $S_{\alpha+1} = C_{\alpha+1} = \alpha + 1$ for each $\alpha < \omega_1$. If $\lim(\alpha)$, let $\langle S_\alpha, C_\alpha \rangle$ be the $<_L$-least pair such that $S_\alpha, C_\alpha \subseteq \alpha$, C_α is closed unbounded in α and $\forall \gamma \in C_\alpha \;\; S_\alpha \cap \gamma \neq S_\gamma$; if no such pair exists, set $S_\alpha = C_\alpha = \alpha$.

Now, suppose that $\langle S_\alpha \mid \alpha < \omega_1 \rangle$ does not satisfy $\diamondsuit$. Then we can find a pair $\langle X, C \rangle$ such that $X, C \subseteq \omega_1$, C is closed and unbounded, and $\forall \gamma \in C \;\; X \cap \gamma \neq S_\gamma$. Let $\langle X, C \rangle$ be the $<_L$-

least such pair.

Let $M \prec L_{\omega_2}$ be countable. Let π be the collapsing isomorphism of M onto some L_β. Since $\langle S_\alpha \mid \alpha < \omega_1 \rangle$ and $\langle X, C \rangle$ are definable in L_{ω_2}, they are elements of M. Let $\alpha = \omega_1 \cap M$. Just as in the proof of Theorem II.6 we obtain $\pi(\omega_1) = \alpha$, $\pi(\langle\langle S_\nu, C_\nu \rangle \mid \nu < \omega_1 \rangle) = \langle\langle S_\nu, C_\nu \rangle \mid \nu < \alpha \rangle$, and $\pi(\langle X, C \rangle) = \langle X \cap \alpha, C \cap \alpha \rangle$. The sentence

"$\langle X \cap \alpha, C \cap \alpha \rangle$ is the $<_L$-least pair of subsets of α such that $C \cap \alpha$ is closed unbounded in α and $\forall \gamma \in C \cap \alpha \quad X \cap \alpha \cap \gamma \neq S_\gamma$ "

holds in L_β, and hence (by absoluteness) in the real world. Thus, by definition, $\langle X \cap \alpha, C \cap \alpha \rangle = \langle S_\alpha, C_\alpha \rangle$. In particular, $X \cap \alpha = S_\alpha$. But the sentence "$C \cap \alpha$ is closed unbounded in α" holds in L_β, hence in the real world. So α is a limit point of C, and since C is closed, $\alpha \in C$. By the definition of C, therefore, $X \cap \alpha \neq S_\alpha$, a contradiction! ∎

In fact, the proof can be strengthened so as to yield: $\diamondsuit$ holds if $V = L[A]$ for some $A \subseteq \omega_1$. A somewhat more ingenious proof gives us that $\diamondsuit^+$ also holds in L. (See [De1].)

We now proceed to show that $\diamondsuit$ may replace the condensation arguments in the construction of a Souslin tree.

<u>Theorem 5</u>

Assume $\diamondsuit$. Then there exists a Souslin tree.

Proof: Let $\langle S_\alpha \mid \alpha < \omega_1 \rangle$ be as in $\diamondsuit$ (in the form given by Theorem 2). By induction we shall construct the levels T_α of a normal tree $T = \langle T, \leq_T \rangle$ of length ω_1 such that $T = \omega_1$ (i.e. its points are the countable ordinals).

(i) $T_0 = \{0\}$.

(ii) Let T_α be given. The elements of T_α are countable ordinals, and hence have a canonical well-ordering. Proceeding along T_α under this ordering, appoint to each $x \in T_\alpha$ the next two ordinals not already used as immediate successors of x . This defines $T_{\alpha+1}$.

(iii) Assume $T|\alpha$ is given for $\lim(\alpha)$. To every $x \in T|\alpha$ choose an α - branch b_x going through x such that $b_x \neq b_y$ when $x \neq y$ and: If S_α is a maximal antichain of $T|\alpha$, then $b_x \cap S_\alpha \neq \emptyset$. (This is clearly possible. We note here that we could have used the original version of $\diamondsuit$, requiring that $b_x \cap A \neq \emptyset$ whenever $A \in S_\alpha$ happens to be a maximal antichain of $T|\alpha$.) The well-ordering of the points in $T|\alpha$ induces a well-ordering of the branches b_x . So we may inductively let the least ordinal not already used be the point extending b_x . This defines T_α, and the construction of T is complete.

By induction it is clear that T is a normal tree of length ω_1 . We now prove that T is Souslin. Let X be a maximal antichain in T, and let

$$A = \{\alpha < \omega_1 \mid \lim(\alpha) \wedge X \cap \alpha \text{ is a maximal antichain of } T|\alpha\}.$$

<u>Claim</u>: A is closed unbounded in ω_1 .

Proof: Suppose $\lim(\alpha)$ and that $A \cap \alpha$ is unbounded in α . We show that $\alpha \in A$. Let $\langle \beta_\nu \mid \nu < \gamma \rangle$ be the monotone enumeration of $A \cap \alpha$. If $X \cap \alpha$ is not a maximal antichain of $T|\alpha$, then there exists an $x \in T|\alpha$ which is incomparable with $X \cap \alpha$. But $x \in T|\beta_\nu$ for some $\nu < \gamma$, so, à fortiori, x is incomparable with $X \cap \beta_\nu$; but this is impossible since $\beta_\nu \in A$.

Hence A is closed.

Next we show that A is unbounded. Note that, by construction, the set $\{\alpha \mid T|\alpha = \alpha\}$ is a (closed and) unbounded subset of ω_1. For $\alpha < \omega_1$ we define α^* as follows. For $x \in T|\alpha$, let y_x be the least element in X comparable with x. Let $Y_\alpha = \{y_x \mid x \in T|\alpha\}$. Set α^* = the least ordinal $> \alpha$ such that $Y_\alpha \subseteq T|\alpha^*$ and $T|\alpha^* = \alpha^*$. Now let $\beta_0 < \omega_1$ be arbitrary. Define inductively $\beta_{n+1} = \beta_n^*$ for $n \in \omega$. Let $\alpha = \sup_{n\in\omega} \beta_n$. We claim that $\alpha \in A$. Let $x \in T|\alpha$. Then $x \in T|\beta_n$ for some $n \in \omega$. Now $y_x \in X \cap T|\beta_{n+1} \subseteq X \cap T|\alpha \subseteq X \cap \alpha$, so x is comparable with an element of $X \cap \alpha$. Hence $\alpha > \beta_0$ and $\alpha \in A$. The proof of the claim is complete.

Now we use $\diamondsuit$. Since $\{\alpha \mid X \cap \alpha = S_\alpha\}$ is stationary, pick $\alpha \in A$ such that $X \cap \alpha = S_\alpha$. Then S_α is a maximal antichain in $T|\alpha$. By construction, every point in T_α lies above a point of S_α. Hence S_α is maximal also in T. But then X cannot contain any more elements than S_α, so $X = S_\alpha \subseteq \alpha$, and hence X is countable. ▮

3. $\diamondsuit$ obtained by forcing

In this section we show that every countable model of ZFC can be extended to a model of $\diamondsuit$ (and hence of $\neg$SH). The similar result for $\diamondsuit^*$ is also valid and will be shown in Chapter VIII, but we prove the result for $\diamondsuit$ separately since the extension is so simple: the $\diamondsuit$-sequence can be defined from any standard generic subset of ω_1.

Theorem 6

Let M be a c.t.m. of ZFC. In M, let $\mathbb{P}$ be the poset of all maps from a countable ordinal into 2, ordered by inclusion (i.e. $\mathbb{P}$ is the standard poset for adding a generic subset of ω_1). Let G be

M-generic for $\mathbb{P}$. Then:

(i) $\mathcal{P}^{M[G]}(\omega) = \mathcal{P}^{M}(\omega)$ and $\omega_1^{M[G]} = \omega_1^{M}$,

(ii) if $M \models 2^{\omega} = \omega_1$, then M and $M[G]$ have the same cardinals,

(iii) $M[G] \models \diamondsuit$.

Proof: $\mathbb{P}$ is clearly ω_1-closed, so (i) is clear. Furthermore, $\mathbb{P}$ trivially satisfies $(2^{\omega})^{+}$ - c.c. , so if $2^{\omega} = \omega_1$, then ω_2 and all larger cardinals are also absolute. So it remains to prove (iii).

We define the sequence $\langle S_\alpha \mid \alpha < \omega_1 \rangle$ from G as follows. If α is not a limit ordinal, set $S_\alpha = \emptyset$. For every limit ordinal α, let j_α be a canonical, $\Sigma_0(M)$ - definable bijection from α onto ω . There exists a unique $p_\alpha \in G$ such that $\operatorname{dom} p_\alpha = \alpha + \omega$. We set for $\alpha < \omega_1$, $\lim(\alpha)$

$$S_\alpha = \{\nu < \alpha \mid p_\alpha(\alpha + j_\alpha(\nu)) = 1\} .$$

We show that $\langle S_\alpha \mid \alpha < \omega_1 \rangle$ satisfies $\diamondsuit$ in $M[G]$. Suppose $X \subseteq \omega_1$ is given, and that $C \subseteq \omega_1$ is any closed unbounded set. Let $p \in G$ be such that

$$p \Vdash \dot{X} \subseteq \check{\omega}_1 \wedge \dot{C} \subseteq \check{\omega}_1 \text{ is closed unbounded.}$$

We claim that $p \Vdash \exists \alpha \in \dot{C}\ (\dot{X} \cap \alpha = \dot{S}_\alpha)$ (and hence that this setence holds in $M[G]$). (Here "$\dot{S}_\alpha$" is an abbreviation for the defining formula in terms of $\dot{G}$.) It is enough to show that $\forall p' \leq p \quad \exists q \leq p' \quad q \Vdash \exists \alpha \in \dot{C}(\dot{X} \cap \alpha = \dot{S}_\alpha)$.

So let $p' \leq p$. Set $\alpha_0 = \operatorname{dom} p'$. Work in M now. By induction, define sequences $\langle \alpha_n \mid n \in \omega \rangle$, $\langle \beta_n \mid n \in \omega \rangle$ of ordinals and sequences $\langle p_n \mid n \in \omega \rangle$, $\langle q_n \mid n \in \omega \rangle$ of conditions such that $p_0 \leq p'$ and for each $n \in \omega$

(a) $\forall\nu < \alpha_n \ (p_n \Vdash \check{\nu} \in \dot{X}$ or $p_n \Vdash \check{\nu} \notin \dot{X})$,

(b) $\forall\nu < \alpha_n \ (p_n \Vdash \check{\nu} \in \dot{C}$ or $p_n \Vdash \check{\nu} \notin \dot{C})$,

(c) $\beta_n > \alpha_n \wedge q_n \leq p_n \wedge q_n \Vdash \check{\beta}_n \in \dot{C} \wedge \operatorname{dom} q_n \geq \beta_n$,

(d) $\alpha_{n+1} = \operatorname{dom} q_n$,

(e) $p_{n+1} \leq q_n$.

(a) and (b) can be satisfied since, by ω_1-closedness, we can easily arrange for countably many statements to be decided by one condition. (c) is also all right, since every extension of p forces C to be unbounded in ω_1 .

Let $\alpha = \sup_{n\in\omega}\alpha_n$ and let $p_\omega = \bigcup_{n\in\omega} p_n$. Then, setting $X_\alpha = \{\nu < \alpha \mid p_\omega \Vdash \check{\nu} \in \dot{X}\}$ and $C_\alpha = \{\nu < \alpha \mid p_\omega \Vdash \check{\nu} \in \dot{C}\}$, clearly $p_\omega \Vdash$ "$\dot{X} \cap \check{\alpha} = \check{X}_\alpha \wedge \dot{C} \cap \check{\alpha} = \check{C}_\alpha$" . Since $p_\omega \Vdash$ "$\dot{C}$ is closed", condition (c) guarantees that $p_\omega \Vdash \check{\alpha} \in \dot{C}$. By (d), $\operatorname{dom} p_\omega = \alpha$. Define $q \leq p_\omega$ by: $\operatorname{dom} q = \alpha + \omega$, $q \restriction \alpha = p$, and $q(\alpha+n) = 1$ iff $j_\alpha^{-1}(n) \in X_\alpha$. Then we readily get that $q \Vdash \dot{X} \cap \check{\alpha} = \dot{S}_{\check{\alpha}}$ (using only the facts that $r \Vdash \check{q} \in \dot{G}$ iff $r \leq q$, and that forcing is absolute with respect to Σ_0-sentences from M). Hence we are done, since $q \leq p'$ and $q \Vdash \exists\alpha \in \dot{C} \ (\dot{X} \cap \alpha = \dot{S}_\alpha)$.

∎

Chapter IV

HOMOGENEOUS SOUSLIN TREES AND LINES

1. A homogeneous Souslin tree

An automorphism of a poset $X = \langle X, \leq \rangle$ is a bijective map $\sigma : X \to X$ such that for all $x,y \in X$, $x \leq y$ iff $\sigma(x) \leq \sigma(y)$ (i.e. an isomorphism from X to X, or a map such that both σ and σ^{-1} embed X in X). In this and the next chapter we will concentrate our interest on automorphism properties of certain Souslin trees, constructed from $\diamondsuit$. Clearly, if T is a tree and σ is an automorphism on T, then for all $x \in T$, $ht(\sigma(x)) = ht(x)$. So let us call a tree T homogeneous if for all $x,y \in T$ with $ht(x) = ht(y)$ there exists an automorphism σ which interchanges x and y (i.e. $\sigma(x) = y$ and $\sigma(y) = x$).

For a moment, let the requirement (iii) in the definition of a normal tree be strengthened so that the number of immediate successors of any non-maximal point be fixed (for instance ω, or 2). It can easily be proved that any two countable normal trees T, T' of the same length α are isomorphic. (Let $\langle b_n \mid n \in \omega \rangle$ be a sequence of α-branches such that $\bigcup_{n\in\omega} b_n = T$, and $\langle d_n \mid n \in \omega \rangle$ similarly for T'. Define a correspondence between the branches b_n and d_m by induction on n in a structure preserving and minimal way. This will give an isomorphism $T \leftrightarrow T'$. (A detailed proof may be found in [Ku 1] p.102.)) By this fact, it is easy to see that if T is a normal tree of length ω_1, then

i) $T|\alpha$ is homogeneous for all $\alpha < \omega_1$,

ii) if $\alpha < \beta < \omega_1$, then any automorphism on $T|\alpha+1$ can be extended to an automorphism on $T|\beta$.

Hence we may be tempted to claim that every normal tree of length ω_1 is homogeneous (or, equivalently, that any two normal trees of length

ω_1 are isomorphic). This is not so, however, as the next chapter will show. So, in order to construct a homogeneous Souslin tree, we need some strategy for what to do at limit ordinals.

Theorem 1

Assume $\diamondsuit$. Then there is a homogeneous Souslin tree.

Proof: Let $\langle S_\alpha \mid \alpha<\omega_1\rangle$ be the sequence given by $\diamondsuit$.
Since $\diamondsuit$ implies CH, H_{ω_1} (the set of all hereditarily countable sets) has cardinality ω_1, and can be "coded" by some $A \subseteq \omega_1$. Hence $L_{\omega_1}[A] = H_{\omega_1}$, and $\omega_1^{L[A]} = \omega_1$. By induction on $\alpha < \omega_1$, define δ_α to be the least ordinal $\delta > \alpha$ such that

(i) $L_\delta[A] \prec H_{\omega_1}$,

(ii) $S_\alpha, \langle \delta_\nu \mid \nu<\alpha\rangle \in L_\delta[A]$.

Set $M_\alpha = L_{\delta_\alpha}[A] = L_{\delta_\alpha}[A \cap \delta_\alpha]$. Then M_α is countable for $\alpha < \omega_1$. We now prove a useful general lemma.

Lemma 2

Let T be a normal tree of length ω_1 such that $x \in T \rightarrow x \in H_{\omega_1}$, and let $C \subseteq \omega_1$ be a closed unbounded set of limit ordinals such that for all $\alpha \in C$: If $T|\alpha \in M_\alpha$ and $x \in T_\alpha$, then $\{y \mid y <_T x\}$ is M_α-generic for $T|\alpha$.

Then T is a Souslin tree.

Proof: Let X be a maximal antichain in T. We will show that X is countable. Let $B \subseteq \omega_1$ be such that $C, X, T, A \in L[B]$. Set $N^* = \langle L_{\omega_2^{L[B]}}[B], \in\rangle$. Define models $N_\nu \prec N^*$ for $\nu < \omega_1$ as follows:

N_0 = the smallest $N \prec N^*$ such that $B \in N$,

$$N_{\alpha+1} = \text{the smallest } N \prec N^* \text{ such that } N_\alpha \cup \{N_\alpha\} \subseteq N ,$$

$$N_\alpha = \bigcup_{\nu<\alpha} N_\nu \text{ for } \lim(\alpha) .$$

Let α_ν be the least ordinal not in N_ν, and let $\pi_\nu : N_\nu \xrightarrow{\sim} \bar{N}_\nu$ be the collapsing map (with $\bigcup \bar{N}_\nu \subseteq \bar{N}_\nu$) for $\nu < \omega_1$. Then for every $\nu < \omega_1$, there is β_ν such that $\bar{N}_\nu = L_{\beta_\nu}[\pi_\nu(B)] = L_{\beta_\nu}[B \cap \alpha_\nu]$. We note that $\pi_\nu \restriction L_{\alpha_\nu}[B \cap \alpha_\nu] = \mathrm{id} \restriction L_{\alpha_\nu}[B \cap \alpha_\nu]$, so if $Y \in N_\nu$ and $Y \subseteq L_{\omega_1}[B]$, then $\pi_\nu(Y) = Y \cap L_{\alpha_\nu}[B \cap \alpha_\nu]$.

The set $\{\alpha_\nu \mid \nu < \omega_1\}$ is obviously closed unbounded in ω_1. By elementary equivalence, $C \cap \alpha_\nu$ is closed unbounded in α_ν, hence $\alpha_\nu \in C$ for $\nu < \omega_1$.

By $\diamondsuit$ there exists an ordinal α such that $\alpha_\alpha = \alpha$ and $B \cap \alpha = S_\alpha$. Since $S_\alpha \in M_\alpha$, $B \cap \alpha \in M_\alpha$. (This is the only point we use $\diamondsuit$ in the whole proof. Note that this application of $\diamondsuit$ is more indirect than earlier.)

<u>Claim:</u> $\beta_\alpha \in M_\alpha$.

Proof: We split the proof into two cases. First, assume $\omega_1^{L[B\cap\gamma]} = \omega_1$ for some $\gamma < \omega_1$. Let γ_0 be the least such ordinal. Since γ_0 is definable in N^* and $N_\alpha \prec N^*$, $\gamma_0 \in N_\alpha$. Hence $\gamma_0 < \alpha$, so $\omega_1^{L[B\cap\alpha]} = \omega_1$. Let $\tau < \omega_1$ be the least ordinal such that α is countable in $L_\tau[B \cap \alpha]$. τ is H_{ω_1}-definable in the parameter $B \cap \alpha$. From the fact that $B \cap \alpha \in M_\alpha \prec H_{\omega_1}$ it follows that $\tau \in M_\alpha$. But since α is uncountable in $L_{\beta_\alpha}[B \cap \alpha]$, we obtain $\beta_\alpha < \tau$, so $\beta_\alpha \in M_\alpha$.

Secondly, assume $\omega_1^{L[B\cap\gamma]} < \omega_1$ for all $\gamma < \omega_1$. If for some $\gamma < \omega_1$, $\omega_2^{L[B\cap\gamma]} = \omega_1$, put $\zeta = \omega_1^{L[B\cap\gamma]}$, whence ζ is countable in $L[B\cap\eta]$ for sufficiently large η, so $\omega_1^{L[B\cap\eta]} = \omega_1$, impossible! Thus $\omega_2^{L[B\cap\alpha]}$ is countable, and hence is H_{ω_1}-definable in the parameter $B \cap \alpha$. So $\omega_2^{L[B\cap\alpha]} \in M_\alpha$. By elementary equivalence, α (the image of ω_1) is mappable onto each $\theta < \beta_\alpha$ in $L_{\beta_\alpha}[B \cap \alpha]$. This means that $\beta_\alpha \leq \omega_2^{L[B\cap\alpha]}$, so $\beta_\alpha \in M_\alpha$ also in this case, and the claim is proved.

Then, recalling that $B \cap \alpha \in M_\alpha = L_{\delta_\alpha}[A]$, we have that $\bar{N}_\alpha = L_{\beta_\alpha}[B \cap \alpha] \subseteq L_{\delta_\alpha}[B \cap \alpha] \subseteq M_\alpha$. So $\pi_\alpha(T) = T|\alpha \in \bar{N}_\alpha \subseteq M_\alpha$, and $\pi_\alpha(X) = X \cap T|\alpha \in \bar{N}_\alpha \subseteq M_\alpha$. By elementary equivalence, $X \cap T|\alpha$ is a maximal antichain in $T|\alpha$.

For $x \in T_\alpha$, the branch $\{y \mid y <_T x\}$ is M_α-generic for $T|\alpha$; hence it intersects $X \cap T|\alpha$, so x lies above a point in $X \cap T|\alpha$. This means that $X \cap T|\alpha$ is a maximal antichain in $T|\alpha{+}1$, and hence also in T. Thus $X = X \cap T|\alpha$, so X is countable. Q.E.D. (Lemma 2)

We turn to the construction of the homogeneous Souslin tree T. T is defined by induction on the levels T_α. The points of T shall be countable 0,1 - sequences s such that $ht(s) = dom(s)$, and the ordering on T will be ordinary inclusion. By induction, it will follow that $T|\alpha \in M_\alpha$ for every $\alpha < \omega_1$.

As usual, $T_0 = \{\emptyset\}$, and if $\alpha = \beta + 1$ for some β, $T_\alpha = \{s^\frown\langle i\rangle \mid s \in T_\beta \wedge i \in 2\}$. Now assume $\lim(\alpha)$. Let b be the least (in $L[A]$) M_α-generic branch of $T|\alpha$. Define $T_\alpha = \{\bigcup d \mid d$ is an α-branch through $T|\alpha$, and $\bigcup d$ and $\bigcup b$ differ only on an initial segment$\}$. Then $T = \bigcup_{\alpha<\omega_1} T_\alpha$ is clearly a normal tree of length ω_1. We show that T is homogeneous.

<u>Claim 1</u>: For each pair $s, s' \in T$ such that $ht(s) = ht(s')$, the set $\{\nu \mid s_\nu \neq s'_\nu\}$ is finite.

Proof: By induction on $\alpha = ht(s) = ht(s')$. Trivial for 0 and successor α . For $\lim(\alpha)$, $s,s' \in T_\alpha$, we know that s and s' are equal except for an initial segment, on which the induction hypothesis holds. So Claim 1 is clear.

Now, for finite $a \subseteq \omega_1$ define $\sigma_a : 2^{<\omega_1} \to 2^{<\omega_1}$ by: $\mathrm{ht}(\sigma_a(s)) = \mathrm{ht}(s)$ and

$$\sigma_a(s)_\nu = \begin{cases} s_\nu & \text{if } \nu \notin a \\ (1-s_\nu) & \text{if } \nu \in a . \end{cases}$$

(So σ_a changes s at finitely many places. Note that $2^{<\omega_1}$ means $\bigcup_{\alpha<\omega_1} 2^\alpha$.)

<u>Claim 2</u>: T is closed under the maps σ_a (for $a \subseteq \omega_1$, $|a| < \omega$).

This follows directly from the definition of T . Claim 1 and 2 immediately imply that T is homogeneous.
Finally, we note that for $\lim(\alpha)$, each α-branch which is extended in T is, in fact, M_α-generic for $T|\alpha$. (For let $b = \sigma_a''(d)$ with d arbitrary. If D is dense in $T|\alpha$ and $D \cap d = \emptyset$, then $\sigma_a''(D) \cap b = \emptyset$, which is impossible since $\sigma_a''(D)$ is dense.) Hence Lemma 2 gives that T is Souslin. ∎

<u>Theorem 3</u>
Assume $\diamondsuit$. Then there is a Souslin tree with exactly ω_1 automorphisms.

Proof: We show that the homogeneous Souslin tree T constructed above has exactly ω_1 automorphisms. Of course, $\sigma_{\{\alpha\}} \restriction T \neq \sigma_{\{\beta\}} \restriction T$ for $\alpha, \beta < \omega_1$, $\alpha \neq \beta$. So T has at least ω_1 automorphisms. The converse direction is an immediate consquence of CH and the following claim.

<u>Claim</u>: If σ is automorphism of T , then there is an $\alpha < \omega_1$

such that for all $x \in T$ and $\nu < \omega_1$, if $\alpha \leq \nu < ht(x)$, then $\sigma(x)_\nu = x_\nu$.

Proof: The claim may equivalently be stated as :-

If σ is an automorphism of T, then

(*) $D = \{y \mid \forall z > y \; \forall \nu < \omega_1 (ht(y) \leq \nu < ht(z) \rightarrow z_\nu = \sigma(z)_\nu)\}$

is dense in T (i.e. $\forall x \in T \; \exists y \geq x \; y \in D$).

Clearly, the claim implies (*), since then $D \supseteq \{y \mid ht(y) > \alpha\}$, which is dense in T.

Conversely, assume D is dense in T. Then $X =_{df} \{x \in D \mid \neg \exists y \in D \; y < x\}$ is a maximal antichain in T. Since T is Souslin, $X \subseteq T|\alpha$ for some $\alpha < \omega_1$. This α satisfies the claim.

Suppose (*) is false. Then there exist an automorphism σ and $x_0 \in T$ such that $\forall y \geq x_0 \; \exists z > y \; \exists \nu < \omega_1 \; (ht(y) \leq \nu < ht(z) \wedge z_\nu \neq \sigma(z)_\nu)$. Define for $y \in T$,

$$D_y = \{z \mid (z > y \wedge \exists \nu (ht(y) \leq \nu < ht(z) \wedge z_\nu \neq \sigma(z)_\nu)) \vee (z | y)\} .$$

Then D_y is dense in T for $y \geq x_0$.

We now proceed as above, letting $N^* = L_{\omega_2^{L[B]}}[B]$, where we now require $B \subseteq \omega_1$ and $T, \sigma, A \in L[B]$. N_0 is the smallest elementary submodel of N^* with $B \in N_0$, $N_{\nu+1}$ is the smallest elementary submodel of N^* containing $N_\nu \cup \{N_\nu\}$, and $N_\lambda = \bigcup_{\nu<\lambda} N_\nu$ for $\lim(\lambda)$. As before, let α_ν be the least ordinal not in N_ν, and $\pi_\nu : N_\nu \cong \bar{N}_\nu = L_{\beta_\nu}[B \cap \alpha_\nu]$. There exists $\alpha = \alpha_\alpha$ such that $S_\alpha = B \cap \alpha \in M_\alpha$, and as before $\beta_\alpha \in M_\alpha$, so $\bar{N}_\alpha \subseteq M_\alpha$. Further we get $\pi_\alpha(T) = T|\alpha$ and $\pi_\alpha(D_y) = D_y \cap T|\alpha$ for $ht(y) < \alpha$.

By elementary equivalence there is an $x_0 \in T|\alpha$ such that for all $y \geq x_0$ with $ht(y) < \alpha$, $D_y \cap T|\alpha$ is dense in $T|\alpha$. Now let $x > x_o$, $ht(x) = \alpha$. Then $\{y \mid y < x\}$ is M_α-generic for $T|\alpha$, and thus intersects $D_y \cap T|\alpha \in M_\alpha$ for each $y < x$ with $y \geq x_o$. This means that there exist arbitrarily large $\nu < \alpha$ such that $x_\nu \neq \sigma(x)_\nu$, which contradicts the fact that the set of such ν's is finite! ∎

For a normal tree T of length ω_1, and $B \subseteq \omega_1$, let $T|B$ be the restriction of T to the set $\bigcup_{\alpha\in B} T_\alpha$. It is obvious that if $C \subseteq \omega_1$ is closed unbounded and T is a normal (Souslin) tree, then so is $T|C$.

Now, let T be as before, and let $C \subseteq \omega_1$ be closed unbounded. An easy modification of the proof of Theorem 3 yields: $T|C$ has exactly ω_1 automorphisms.

Finally, a remark about forcing with Souslin trees.
If M is a c.t.m. of ZFC, and T is a normal Souslin tree in M, then of course $M \models$ "T satisfies c.c.c.". But T^2 (i.e. $T \times T$ with the usual product ordering) never satisfies c.c.c. in M. (The proof of this is an easy exercise.) Hence, by forcing with T^2 we may collapse cardinals. (This fact can also be said in another way: By the product lemma, forcing with T^2 is the same as forcing twice with T. So, since T does not satisfy c.c.c. in $M[b]$ with b M-generic for T, cardinals <u>may</u> be collapsed in one further extension.)

We now give an example where we actually <u>must</u> collapse cardinals. Let T be the above defined homogeneous Souslin tree, constructed inside a c.t.m. M of ZFC. Let b be M-generic for T. Define in $M[b]$:

$$T^* = \{s \in 2^{<\omega_1} \mid \{\nu \mid s_\nu = 1\} \text{ is finite}\} .$$

Then $M[b] \models T \cong T^*$. The isomorphism $\sigma \in M[b]$ is given by: $ht(\sigma(s)) = ht(s)$ and $\sigma(s)_\nu = |s_\nu - (\bigcup b)_\nu|$ for $\nu < ht(s)$. If now

d is $M[b]$-generic for T, then $\sigma''d$ is $M[b]$-generic for T^*. Then ω_1^M is countable in $M[b,d]$. For $\bigcup d$ has cofinally many 1's, but restricted to any $\alpha < \omega_1^M$ the number of 1's is finite. Hence ω_1^M is ω-cofinal in $M[b,d]$. So forcing with T^2 actually collapses cardinals in this case.

2. Homogeneous Souslin lines

Recall that an ordered continuum is a complete densely ordered set without end-points, and that a Souslin line S is a non-separable ordered continuum having the Souslin property (i.e. every family of disjoint open intervals in S is countable). Now, let $S = \langle S, \leq \rangle$ be an ordered continuum. We say that $S = \langle S, \leq \rangle$ is <u>homogeneous</u> if to any four points x_0, x_1, y_0, y_1 with $x_0 < x_1$ and $y_0 < y_1$ there exists an automorphism σ on S such that $\sigma(x_i) = y_i$ for $i = 1,2$. We say that S is <u>reversible</u> if $\langle S, \leq \rangle$ is isomorphic to $\langle S, \geq \rangle$ (i.e. S with the reverse ordering).

If T is a normal Aronszajn tree, we say that T is a <u>standard tree</u> if i) $s \in T_\alpha$ implies $s \in 2^\alpha$ (i.e. s is an α-sequence from $\{0,1\}$), ii) T is ordered by inclusion, and iii) if $s \in T_\alpha$ and $\lim(\alpha)$, then s does not end in a 0-sequence or a 1-sequence (i.e. for all $\beta < \alpha$ there exists $\nu_0, \nu_1 \geq \beta$ such that $s_{\nu_0} = 0$ and $s_{\nu_0} = 1$). We now describe a general method for obtaining an ordered continuum $S = S_T$ from a standard tree T.

Let T be a standard tree. Let $S' = S'_T$ be the set of all $b \in 2^{<\omega_1}$ such that $\{b \restriction \nu \mid \nu < \operatorname{dom} b\}$ is a maximal branch through T. S' is not densely ordered. For, if $s \in T$, let $s_0 = s^\frown\langle 0,1,1\,(\omega \text{ times})\ldots\rangle$ and $s_1 = s^\frown\langle 1,0,0,\ldots\rangle$. Then $s_0, s_1 \in S'$ and there is no point in S' between s_0 and s_1. Conversely, any "gap" of S' is clearly of this type. Thus let $S = S_T$ be obtained by removing from S' all the points s_0 for $s \in T$ together with the end-points $\langle 0,0,\ldots \rangle$ and $\langle 1,1,\ldots \rangle$. Then S is an ordered continuum. For if $A \subseteq S$ has an upper bound, define $b \in S$ by induction such that $b_\nu = \max\{x_\nu \mid x \in A \wedge b\restriction\nu \subsetneqq x\}$ if such an x exists, undefined otherwise. Then $b = \operatorname{lub} A$.

For $s \in T$, let

$$I_s = \{x \in S \mid s \subset x\} .$$

Then every open interval of S contains an I_s. Hence, as in the proof of Theorem II.4, one easily concludes that S is not separable, and if T is Souslin, then S has the Souslin property, and is thus a Souslin line.

For convenience, let us call an automorphism σ on a subset of a standard tree T <u>lexicographical</u> if σ preserves the lexicographical ordering of the points in T. (The lexicographical ordering, $<_1$, is defined by $s <_1 s'$ iff $\exists \beta < \omega_1 (s\restriction\beta = s'\restriction\beta \wedge s_\beta < s'_\beta)$.) σ is <u>antilexicographical</u> if σ reverses the lexicographical ordering of T. (Note that this has nothing to do with the partial ordering $\subseteq$ of T itself.)

<u>Theorem 4</u>

Assume $\diamondsuit$. Then there exists a Souslin line S such that

(i) S is homogeneous,

(ii) S has exactly ω_1 automorphisms,

(iii) S is not reversible.

Proof: Let T be the homogeneous Souslin tree (with exactly ω_1 automorphisms) constructed in the previous section. T is clearly standard, so let $S = S_T$ be the Souslin line obtained from T as just described. We now prove some central lemmas concerning the relations between automorphisms on T and S. In fact, Lemma 5 and 7 are valid for arbitrary standard trees T, and

Lemma 6 and 8 for arbitrary standard Souslin trees.

Lemma 5

Let $C \subseteq \omega_1$ be closed unbounded. If σ is a lexicographical automorphism on $T|C = \{s \in T \mid ht(s) \in C\}$, then there exists a unique automorphism $\bar{\sigma}$ on S so that $\bar{\sigma}'' I_s = I_{\sigma(s)}$ for all $s \in T|C$.

Proof: Without loss of generality we may assume that $\alpha \in C \rightarrow \lim(\alpha)$. (For if the lemma is true for the set $\{\alpha + \omega \mid \alpha \in C\}$, it is also true for C.) For $\alpha \in C$, set

$$S_\alpha = \{b \upharpoonright \alpha \mid b \in S\} .$$

Then each S_α is an ordered continuum, and T_α is dense in S_α. (In fact, by Theorem II.1, S_α is isomorphic to $\mathbb{R}$. We also note that if we set $I_s = \{s\}$ for $s \in S$, then $S = \bigcup_{s \in S_\alpha} I_s$.) Hence there is a unique automorphism $\bar{\sigma}_\alpha$ on S_α extending $\sigma \upharpoonright T_\alpha$, for $\alpha < \omega_1$.

We claim that for $\alpha < \beta$, $\bar{\sigma}_\alpha \upharpoonright (S_\alpha - T_\alpha) \subseteq \bar{\sigma}_\beta \upharpoonright (S_\beta - T_\beta)$. For let $s \in S_\alpha - T_\alpha$ such that $s = \text{lub}\{s_i \mid i \in \omega\}$ with each $s_i \in T_\alpha$. Then $s = \text{lub}\{s_i' \mid i \in \omega\}$ whenever each $s_i' \in T_\beta$, $s_i \subset s_i'$. Hence $\bar{\sigma}_\alpha(s) = \text{lub}\{\sigma_\alpha(s_i) \mid i \in \omega\} = \text{lub}\{\sigma_\beta(s_i') \mid i \in \omega\} = \bar{\sigma}_\beta(s)$.

Hence, by setting

$$\bar{\sigma} = \bigcup_{\alpha \in C} \bar{\sigma}_\alpha | (S_\alpha - T_\alpha) ,$$

we obtain an automorphism on S. The uniqueness of $\bar{\sigma}$ and the fact that $\bar{\sigma}''(I_s) = I_{\sigma(s)}$ for $s \in T|C$ are trivially verified.

Q.E.D. (Lemma 5).

We are now in a position to prove that $S = S_T$ is homogeneous. Note that, by homogeneity, T has the following property:

(*) Let $s, s' \in T$ with $ht(s) \geq ht(s')$. Define s^* by: $ht(s^*) = ht(s)$, $s^*_\nu = s'_\nu$ for $\nu < ht(s')$, otherwise $s^*_\nu = s_\nu$. Then $s^* \in T$.

Let $a_0, a_1, b_0, b_1 \in S$ with $a_0 < a_1$ and $b_0 < b_1$. Let α be a limit ordinal such that $a_i, b_i \in S_\alpha$ $(i = 1,2)$. Let $R_0 = T_\alpha \cup \{a_0, a_1\}$, $R_1 = T_\alpha \cup \{b_0, b_1\}$. Since both R_0 and R_1 are countable, densely ordered sets, there is an order-preserving $\sigma : R_0 \xrightarrow{\sim} R_1$ such that $\sigma(a_i) = b_i$ $(i = 1,2)$. Define $\sigma^* : T|(\omega_1 - \alpha) \xrightarrow{\sim} T|(\omega_1 - \alpha)$ by

$$\sigma^*(s)_\nu = \begin{cases} \sigma(s \restriction \alpha)_\nu & \text{for } \nu < \alpha \\ s_\nu & \text{otherwise.} \end{cases}$$

Then σ^* is lexicographical on $T|(\omega_1 - \alpha)$. By Lemma 5 we may define $\bar{\sigma} : S \xrightarrow{\sim} S$ by $\bar{\sigma}'' I_s = I_{\sigma^*(s)}$ for $s \in T|(\omega_1 - \alpha)$. Then it follows readily that $\bar{\sigma}(a_i) = b_i$ for $i = 1,2$. Hence we have proved (i) in Theorem 4. To prove (ii), we need the "converse" of Lemma 5.

<u>Lemma 6</u>

Let σ be an automorphism on S. Then there exists a closed unbounded $C \subseteq \omega_1$ such that if $s \in T_\alpha$ and $\alpha \in C$, then $\sigma'' I_s = I_{s'}$ for some $s' \in T_\alpha$. Hence if we define $\tilde{\sigma}$ by $\tilde{\sigma}(s) =$ that s' such that $\sigma'' I_s = I_{s'}$, then $\tilde{\sigma}$ is a lexicographical automorphism on $T|C$.

Proof: For $s \in T$, let $X_s = \{x \in T \mid I_x \subseteq \sigma'' I_s\}$, and let $Y_s = \{y \in X_s \mid \neg \exists z \in X_s \;\; z \subsetneqq y\}$. Since $\sigma'' I_s$ is an interval, it is clear that $\sigma'' I_s = \bigcup_{y \in Y_s} I_y$. Now, each Y_s is an antichain, hence countable.

Set

$$A = \{\alpha < \omega_1 \mid \lim(\alpha) \wedge \forall s \in T|\alpha \;\; Y_s \subseteq T|\alpha\}.$$

Clearly, A is closed unbounded. We claim that if $\alpha \in A$ and

$s \in T_\alpha$, then there is an $s' \in T_\alpha$ such that $I_{s'} \subseteq \sigma'' I_s$.

Choose $s' \in T_\alpha$ such that $I_{s'} \cap \sigma'' I_s \neq \emptyset$. Then for every $x \subset s$ there exists a $y < s'$ such that $y \in Y_x$, i.e. $I_y \subseteq \sigma'' I_x$. Hence $I_{s'} = \bigcap_{y \subset s'} I_y \subseteq \bigcap_{x \subset s} \sigma'' I_x = \sigma''(\bigcap_{x \subset s} I_x) = \sigma'' I_s$.

Analogously we can find a closed unbounded $B \subseteq \omega_1$ such that if $\alpha \in B$ and $s \in T_\alpha$, then there is an $s' \in T_\alpha$ such that $I_{s'} \subseteq \sigma^{-1}{}'' I_s$.

Let $C = A \cap B$. Let $s \in T_\alpha$ where $\alpha \in C$. Then there are $s', s'' \in T_\alpha$ such that $I_{s'} \subseteq \sigma'' I_s$ and $I_{s''} \subseteq \sigma^{-1}{}'' I_{s'}$. Then if $b \in I_{s''}$, then $\sigma(b) \in I_{s'}$, so $b \in I_s$. Hence $b \supset s''$ implies $b \supset s$, so we must have $s'' = s$. Thus $I_{s''} = I_s$, so $I_{s'} = \sigma'' I_s$.

Q.E.D. (Lemma 6).

(ii) in the theorem now follows from the

<u>Claim</u>: If σ is an automorphism on S, then there is an $\alpha < \omega_1$ such that if $b \in S$ and $\alpha \leq \nu < \operatorname{dom} b$, then $\sigma(b)_\nu = b_\nu$.

Proof: Let σ be an automorphism on S, and let $\tilde{\sigma} : T|C \xrightarrow{\sim} T|C$ be as in Lemma 6, where C is closed unbounded. By the proof of Theorem 3 there is an $\alpha < \omega_1$ such that for all $s \in T|C$, if $\alpha \leq \nu < ht(s)$, then $\tilde{\sigma}(s)_\nu = s_\nu$. The claim is immediate.

With trivial changes, the proofs of the two preceeding lemmas yield the following

<u>Lemma 7</u>

Let $C \subseteq \omega_1$ be closed unbounded. If σ is an antilexicographical automorphism on $T|C$, then there is a unique isomorphism $\tilde{\sigma} : \langle S, \leq \rangle \tilde{\leftrightarrow} \langle S, \geq \rangle$ such that $\tilde{\sigma}'' I_s = I_{\sigma(s)}$ for all $s \in T|C$.

Lemma 8

Let $\sigma : \langle S, \leq \rangle \tilde{\leftrightarrow} \langle S, \geq \rangle$. Then there exists a closed unbounded $C \subseteq \omega_1$ such that if $s \in T_\alpha$ and $\alpha \in C$, there is an $s' \in T_\alpha$ with $\sigma'' I_s = I_{s'}$. Hence $\tilde{\sigma}$ is an antilexicographical automorphism on $T|C$ if we set $\tilde{\sigma}(s)$ = that s' such that $\sigma'' I_s = I_{s'}$.

Now, assume S is reversible, and let $\sigma : \langle S, \leq \rangle \tilde{\leftrightarrow} \langle S, \geq \rangle$. By Lemma 8 there exists a closed unbounded $C \subseteq \omega_1$ such that $\tilde{\sigma} : T|C \tilde{\leftrightarrow} T|C$, and $\tilde{\sigma}$ reverses the lexicographical ordering $<_1$ on $T|C$. On the other hand, by the proof of Theorem 3, there is $\alpha < \omega_1$ such that if $\alpha \leq \nu < ht(s)$, then $\tilde{\sigma}(s)_\nu = s_\nu$. Thus let $s, s' \in T|A$ such that $s <_1 s'$ and $s \upharpoonright \alpha = s' \upharpoonright \alpha$. Then $\tilde{\sigma}(s) <_1 \tilde{\sigma}(s')$, contradiction! The proof of Theorem 4 is complete. ▮

Theorem 9

Assume $\diamondsuit$. Then there exists a Souslin line S such that

(i) S is homogeneous,

(ii) S has exactly ω_1 automorphisms,

(iii) S is reversible.

Proof: We only have to make some minor changes of the proof of Theorem 1, 3 and 4. For $s \in 2^{<\omega_1}$, let $\bar{s}$ be defined by $\operatorname{dom} \bar{s} = \operatorname{dom} s$ and $\bar{s}_\nu = 1 - s_\nu$ for $\nu < \operatorname{dom} s$. Define T as follows: $T_0 = \{\emptyset\}$, $T_{\alpha+1} = \{s ^\frown \langle i \rangle \mid s \in T_\alpha \text{ and } i \in 2\}$. For $\lim(\alpha)$, let b be M_α-generic for $T|\alpha$ (with M_α as before). Let T_α consist of all branches b' such that $\bigcup b'$ differs from $\bigcup b$ or $\overline{\bigcup b}$ only on an initial segment. It follows that

(1) If $s, s' \in T$ with $ht(s) \geq ht(s')$, then $s^* \in T$, where $ht(s^*) = ht(s)$, $s^*_\nu = s'_\nu$ for $\nu < ht(s')$, and $s^*_\nu = s_\nu$ for $ht(s') \leq \nu < ht(s)$.

From (1) it follows that T is homogeneous. Indeed, let $s, s' \in T$, $ht(s) = ht(s')$. For $x \in T$ let β_x be the largest $\beta \leq ht(s), ht(x)$ such that $x \upharpoonright \beta = s \upharpoonright \beta$ or $x \upharpoonright \beta = s' \upharpoonright \beta$. Define $\sigma : T \overset{\sim}{\leftrightarrow} T$ by

$$\sigma(x)_\nu = \begin{cases} s'_\nu & \text{if } \nu < \beta_x \text{ and } x \upharpoonright \beta_x = s \upharpoonright \beta_x, \\ s_\nu & \text{if } \nu < \beta_x \text{ and } x \upharpoonright \beta_x = s' \upharpoonright \beta_x, \\ x_\nu & \text{otherwise.} \end{cases}$$

Then $\sigma(s) = s'$ and $\sigma(s') = s$.

By Lemma 2, T is Souslin. Furthermore, we have

(2) If $s, s' \in T$, $ht(s) = ht(s') = \alpha$ and $\lim \alpha$, then there is $\nu < \alpha$ such that $s_\iota = s'_\iota$ for $\iota \geq \nu$ or $s_\iota = 1 - s'_\iota$ for $\iota \geq \nu$.

Hence we obtain

(3) If $a \subseteq \omega_1$ is closed unbounded and σ is an automorphism on $T|A$, then there is an $\alpha < \omega_1$ such that either i) $ht(s) > \nu \geq \alpha \rightarrow s_\nu = \sigma(s_\nu)$ or ii) $ht(s) > \nu \geq \alpha \rightarrow s_\nu = 1 - \sigma(s_\nu)$.

Thus T has ω_1 automorphisms.

If we let $S = S_T$, it follows that S is a homogeneous Souslin line with exactly ω_1 automorphisms. But in this case, S is closed under the mapping $s \mapsto \bar{s}$. Hence S is reversible. ▮

Chapter V

RIGID SOUSLIN TREES AND LINES

1. A rigid Souslin tree

A poset $X = \langle X, \leq \rangle$ is called rigid if $\mathrm{id} \restriction X$ is the only automorphism on X. We now prove that in L, rigid Souslin trees exist.

Theorem 1

Assume $\diamondsuit$. Then there exists a rigid Souslin tree.

Proof: Let $A \subseteq \omega_1$ and M_α $(\alpha < \omega_1)$ be as in the proof of Theorem IV.1. We define a standard tree T as follows. $T_0 = \{\emptyset\}$ and $T_{\alpha+1} = \{s^\frown\langle i \rangle \mid s \in T_\alpha \wedge i \in 2\}$ for $\alpha < \omega_1$. Now let $\alpha < \omega$ with $\lim(\alpha)$. As an induction hypothesis, assume $T|\alpha \in M_\alpha$. To define T_α we force over M_α with the set of conditions $\mathbb{P} = \langle \mathbb{P}, \leq_{\mathbb{P}} \rangle \in M_\alpha$, defined as follows:

$$\mathbb{P} = \{p \mid \exists a \subset \omega(|a| < \omega \wedge p : a \to T|\alpha)\},$$

$$p \leq_{\mathbb{P}} q \longleftrightarrow \mathrm{dom}(p) \supseteq \mathrm{dom}(q) \wedge \forall n \in \mathrm{dom}(q) \quad (p_n \supseteq q_n).$$

Let $G \subset \mathbb{P}$ be the least (in $L[A]$) M_α-generic set for $\mathbb{P}$. Since M_α is countable in $M_{\alpha+1}$, $G \in M_{\alpha+1}$. (Hence the hypothesis that $T|\beta \in M_\beta$ for all $\lim(\beta)$ will be trivial to verify.) Define for $n \in \omega$

$$b_n = \{p_n \mid p \in G\}.$$

Claim:

i) Each b_n is an α-branch of $T|\alpha$,

ii) $b_n \neq b_m$ for $n \neq m$,

iii) each b_n is M_α-generic for $T|\alpha$,

iv) if $n_1, \ldots, n_m$ are distinct, then $b_{n_1} \times b_{n_2} \times \ldots \times b_{n_m}$ is M_α-generic for

$(T|\alpha)^m$ (with the product ordering),

v) $T|\alpha = \bigcup_{n\in\omega} b_n$.

Proof: i), ii) and iii) clearly follow from iv). We prove iv). Let $n_1,\ldots,n_m$ be distinct. Let $D \subseteq (T|\alpha)^m$ be dense and closed under extensions. Let

$$D^* = \{p \in \mathbb{P} \mid \langle p_{n_1},\ldots,p_{n_m}\rangle \in D\} .$$

Then D^* is a dense initial segment of $\mathbb{P}$, so let $p \in G \cap D^*$. Then $\langle p_{n_1},\ldots,p_{n_m}\rangle \in b_{n_1} \times \ldots \times b_{n_m} \cap D$.

To prove v) let $s \in T|\alpha$. Define

$$D' = \{p \in \mathbb{P} \mid \exists n \in \mathrm{dom}(p) \quad (p_n \supseteq s)\} .$$

Then D' is clearly a dense initial segment of $\mathbb{P}$. Let $p \in G \cap D'$. Then for some $n \in \mathrm{dom}(p)$, $s \subseteq p_n \in b_n$, so $s \in b_n$, and the claim is proved.

Set $T_{\alpha+1} = \{\bigcup b_n \mid n \in \omega\}$. By v) in the claim, $T|(\alpha+1)$ is still normal. Hence $T = \bigcup_{\alpha<\omega_1} T_\alpha$ is a normal tree of length ω_1.

By iii) of the claim and Lemma IV.2, T is Souslin (and standard).

It remains to prove that T is rigid. Suppose σ is a non-trivial automorphism on T. Let $B \subseteq \omega_1$ be such that $T,\sigma,A \in L[B]$. Set $N^* = L_{\omega_2^{L[B]}}[B]$. Define inductively N_ν as follows: N_0 is the smallest elementary submodel of N^* such that $B \in N_0$; $N_{\nu+1}$ is the smallest elementary submodel of N^* including $N_\nu \cup \{N_\nu\}$, and $N_\alpha = \bigcup_{\nu<\alpha} N_\nu$ for $\lim(\alpha)$. For $\nu < \omega_1$, let α_ν be the least ordinal not in N_ν and $\pi_\nu : N_\nu \xrightarrow{\sim} \bar{N}_\nu = L_{\beta_\nu}[B \cap \alpha_\nu]$ the collapsing map.

As in the proof of Lemma IV.2 there is an $\alpha = \alpha_\alpha$ such that

$\bar{N}_\alpha \subseteq M_\alpha$. We know that $\pi_\alpha(T) = T|\alpha$ and $\pi_\alpha(\sigma) = \sigma\restriction(T|\alpha)$.

By elementary equivalence, $\sigma\restriction(T|\alpha)$ is non-trivial, so let $x_0 \in T|\alpha$ be such that $\sigma(x_0) \neq x_0$. Pick $x \in T_\alpha$ such that $x \supset x_0$. Then $\sigma(x) \neq x$. By the construction of T_α there is $m \neq n$ such that $x = \bigcup b_m$ and $\sigma(x) = \bigcup b_n$. Thus $\sigma(x) = \bigcup_{\nu<\alpha} \sigma(x\restriction\nu) \in M_\alpha[x] = M_\alpha[b_m]$, so $b_n \in M_\alpha[b_m]$.

Now, $b_m \times b_n$ is M_α-generic for $(T|\alpha)^2$ by iv) in the claim. By the product lemma (Lemma I.8), b_n is $M_\alpha[b_m]$-generic for $T|\alpha$, hence $b_n \notin M_\alpha[b_m]$. Contradiction ! ∎

In fact, the rigid Souslin tree just described was the tree which Jensen defined in his first construction of a Souslin tree in L . We also note that the last part of the above proof may easily be modified to give the following result: If T is the above constructed rigid Souslin tree, and $C \subseteq \omega_1$ is closed unbounded, then $T|C$ is also rigid. We will at once use this result to obtain a rigid Souslin line.

2. Rigid Souslin lines

Theorem 2

Assume $\diamondsuit$. Then there exists a Souslin line S such that

(i) S is rigid,

(ii) S is not reversible.

Proof: Let T be the above rigid Souslin tree. Let $S = S_T$ as defined in the beginning of section IV.2. Then S is a Souslin line. Now, let σ be a non-trivial automorphism on S , and let $\sigma(b) \neq b$. By Lemma IV.6 there is a closed unbounded $C \subseteq \omega_1$ and an automorphism $\tilde{\sigma}$ on $T|C$ such that $\sigma'' I_s =$

$I_{\tilde{\sigma}(s)}$ whenever $s \in T|C$. Since $T|C$ is rigid, $\tilde{\sigma}$ is the identity, so $\sigma'' I_s = I_s$ whenever $s \in T|C$.

Let $\alpha \in C$ be such that $\alpha \geq \max\{\mathrm{dom}(b), \mathrm{dom}(\sigma(b))\}$. Then one easily finds a $d \in S$ such that $\sigma(d) \neq d$, $\mathrm{dom}(d) > \alpha$, and $\sigma(d) \notin I_s$, where $s = d \restriction \alpha \in T|C$. But, since $d \in I_s$, $\sigma(d) \in \sigma'' I_s = I_s$, a contradiction. Hence we have proved that S is rigid.

If now S was reversible, then, by Lemma IV.8, there would have been an antilexicographical automorphism $\tilde{\sigma}$ on $T|C$ for some closed unbounded $C \subseteq \omega_1$. But then $\tilde{\sigma} = \mathrm{id} \restriction (T|C)$, which is impossible, since id is not at all antilexicographical!

∎

One might be tempted to conjecture that a rigid ordered continuum cannot be reversible. This is not so, however. We now construct from $\diamondsuit$ a rigid Souslin line which <u>is</u> reversible.

<u>Theorem 3</u>

Assume $\diamondsuit$. Then there exists a Souslin tree T with exactly two automorphisms, namely the identity on T and the mapping $s \curvearrowright \bar{s}$.

Proof: We recall at once that for $s \in 2^{<\omega_1}$, $\bar{s}$ is defined by: $\mathrm{dom}(\bar{s}) = \mathrm{dom}(s)$ and $\bar{s}_\nu = 1 - s_\nu$ for $\nu < \mathrm{dom}(s)$. The tree T is defined as the rigid tree, but with the following modification: If $\lim(\alpha)$, then $\mathbb{P}, G, b_n$ $(n \in \omega)$ are as before. But we now define T_α to be $\{\cup b_n \mid n \in \omega\} \cup \{\overline{\cup b_n} \mid n \in \omega\}$. Of course, every branch $\{\bar{s} \mid s \in b_n\}$ is M_α-generic for T_α, so T is Souslin by Lemma IV.2. It is also clear from the construction that T is closed under the automorphism $s \curvearrowright \bar{s}$.

As in the final part of the proof of Theorem 1, the assumption that σ is different both from the identity and the mapping

$s \rightsquigarrow \bar{s}$ readily leads to a contradiction. For let $\alpha = \alpha_\alpha$ be as before. Then for every $x \in T_\alpha$, $\sigma(x) \in M_\alpha[x]$, hence either $\sigma(x) = x$ or $\sigma(x) = \bar{x}$. Let $x_0 \in T|\alpha$ be such that $x_0 \neq \sigma(x_0) \neq \bar{x}_0$. Then, taking $x \in T_\alpha$, $x \supset x_0$, we have that $\sigma(x) \neq x$, hence $\sigma(x) = \bar{x}$. But then $\sigma(x_0) = \bar{x}_0$, a contradiction! ▮

By modifying the proof, one finds that this theorem is also true with T replaced by $T|C$ whenever $C \subseteq \omega_1$ is closed unbounded.

Theorem 4

Assume $\diamond$. Then there exists a Souslin line S such that

(i) S is rigid,

(ii) S is reversible.

Proof: Let T be the tree of Theorem 3. Set $S = S_T$. From the note above and Lemma IV.6 we get that S is rigid, exactly as in the proof of Theorem 2. Furthermore, the mapping $b \rightsquigarrow \bar{b}$ obviously gives an isomorphism from $\langle S, \leq \rangle$ onto a continuum $\bar{S}$ which is isomorphic to $\langle S, \geq \rangle$. ($\bar{S}$ is obtained by removing the right-hand sides of all gaps from S_T' as defined in Chapter IV, section 2.) ▮

3. Some final remarks

We know from $\diamond$ that there are two non-isomorphic Souslin trees. Indeed, if T is the rigid Souslin tree from section 1, then $T^{\langle 0 \rangle} \not\cong T^{\langle 1 \rangle}$. In fact, by an easy argument, Jech [Je 3] has proved that $\diamond$ implies that there are 2^{ω_1} non-isomorphic Souslin trees. This is analogous to the "classical" result of Gaifman and Specker [Ga Sp] that (in ZFC) there are 2^{ω_1} different isomorphism types of Aronszajn trees. In this context we also mention that Devlin [De 2] has extended a theorem of Baumgartner and proved that, assuming $\diamondsuit$, there are 2^{ω_1} non-isomorphic Aronszajn trees which are $\mathbb{R}$-embeddable but not $\mathbb{Q}$-embeddable.

For normal trees T of length ω_1, let $\sigma(T)$ denote the number of automorphisms on T. Hence, if T is rigid, then $\sigma(T) = 1$, and $\sigma(T) = \omega_1$ for the homogeneous Souslin tree T constructed in Chapter IV. In fact, Jech [Je 3] has proved that if T is a normal tree of length ω_1, then $\sigma(T)$ is either finite, or $2^{\omega} \leq \sigma(T) \leq 2^{\omega_1}$ and $\sigma(T)^{\omega} = \sigma(T)$. He has also proved that it is consistent that there is a Souslin tree T with $\sigma(T)$ of arbitrary prescribed cardinality $\varkappa$ between 2^{ω} and 2^{ω_1} provided $\varkappa^{\omega} = \varkappa$.

Jensen has proved that if we assume $\diamondsuit^+$, then there is a homogeneous Souslin tree T such that $\sigma(T) \geq \omega_2$. (It is doubtful whether this is provable from $\diamondsuit$.) We scetch the proof. Let $\langle S_\alpha \mid \alpha < \omega_1 \rangle$ satisfy $\diamondsuit^+$. Define δ_α, M_α as in the proof of Theorem IV.1. A new hierarchy of models M'_α is defined as follows: $M'_\alpha = L_{\delta'_\alpha}[A]$, where δ'_α is the least δ such that $L_\delta[A] \models ZF^- +$ "δ_α is countable". Thus M_α is countable in M'_α. A tree T is defined as before, except for the case $\lim(\alpha)$: An α-branch b of $T|\alpha$ is extended iff b is M_α-generic for $T|\alpha$ and $b \in M'_\alpha$. It easily follows that T is a normal Souslin tree which is homogeneous. It also follows that if $\sigma \in M_\alpha$ is an automorphism on $T|\alpha$, then σ is extendable, i.e. σ can be extended to an automorphism on $T|(\alpha+1)$. Furthermore, for $\alpha < \beta < \omega_1$ with $\lim(\beta)$, if $\sigma \in M_\beta$ is an extendable automorphism on $T|\alpha$, then $M_\beta \models$ "there are uncountably many automorphisms on $T|\beta$ extending σ". (This is so because $M_\beta \prec H_{\omega_1}$.) Now, let $\langle \sigma_\nu \mid \nu < \omega_1 \rangle$ be a sequence of automorphisms on T. We claim that there is an automorphism $\rho \neq \sigma_\nu$ $(\nu < \omega_1)$, thereby proving the theorem. Let $B \subseteq \omega_1$ code up T, $\langle \sigma_\nu \mid \nu < \omega_1 \rangle$ and A. Let $N^* = L_{\omega_2^{L[B]}}[B]$, $N_\nu \prec N^*$, α_ν, $\bar{N}_\nu$ be as in the proof of Theorem IV.1. By $\diamondsuit^+$ there is a $C \subseteq \omega_1$ such that $\alpha \in C \rightarrow B \cap \alpha, C \cap \alpha \in S_\alpha$. Let $\langle \gamma_\nu \mid \nu < \omega_1 \rangle$ be the monotone enumeration of the closed set $\{\gamma \in C \mid \gamma = \alpha_\gamma\}$. As before we obtain $\bar{N}_{\gamma_\nu} \subset M_{\gamma_\nu}$. By induction on ν we define a sequence $\langle \rho_\nu \mid \nu < \omega_1 \rangle$: ρ_0 is an automorphism on $T|\gamma_0$ in M_{γ_0} different from all $\sigma_\iota \restriction (T|\gamma_0)$ for $\iota < \gamma_0$. $\rho_{\nu+1} \in M_{\gamma_{\nu+1}}$ is an automorphism on $T|\gamma_{\nu+1}$ such that $\rho_{\nu+1} \supset \rho_\nu$ and $\rho_{\nu+1} \neq \sigma_\iota \restriction (T|\gamma_{\nu+1})$ for $\iota < \gamma_{\nu+1}$. $\rho_\lambda = \bigcup_{\nu<\lambda} \rho_\nu$ for $\lim(\lambda)$. It can be proved that $\rho_\lambda \in M_{\gamma_\lambda}$. Now let $\rho = \bigcup_{\nu<\omega_1} \rho_\nu$. Then $\rho \neq \sigma_\nu$ for $\nu < \omega_1$, and we are done.

Since the above constructed tree is closed under the mapping $s \rightsquigarrow \bar{s}$, we get the following result: Assuming $\diamondsuit^+$, there exists a homogeneous, reversible Souslin line which has at least ω_2 automorphisms.

The notion of a homogeneous tree may be strengthened to the following. Let us call a tree T <u>totally homogeneous</u> if T is isomorphic to T^x for all $x \in T$. (Recall that $T^x = \{y \in T \mid x \leq y\}$.) Jensen has proved that $\diamondsuit$ implies that there is a totally homogeneous Souslin tree T with ω_1 automorphisms. T_α is defined as follows for $\lim(\alpha)$: If $\nu + \alpha = \alpha$ for all $\nu < \alpha$, let s^* be the union of a M_α-generic branch. Let $s \in 2^\alpha$ lie in T_α iff there are $s_0, s_1 \in T|\alpha$ and s_2 such that $s_0 {}^\frown s_2 = s^*$ and $s_1 {}^\frown s_2 = s$. If $\alpha = \beta + \gamma$ for some $\beta, \gamma < \alpha$, let T_α consist of all $s_1 {}^\frown s_2$ with $s_1, s_2 \in T|\alpha$ and $ht(s_1) + ht(s_2) = \alpha$.

The last concepts we shall mention are concerned with Boolean algebras. A complete Boolean algebra $\mathbb{B}$ is called <u>homogeneous</u> if to any $b, b' \in \mathbb{B} - \{\mathbb{0}, \mathbb{1}\}$ there exists an automorphism σ such that $\sigma(b) = b'$. It can be proved (if $|\mathbb{B}| > 4$) that $\mathbb{B}$ is homogeneous iff for all $b \in \mathbb{B}$, $\mathbb{B}$ is isomorphic to $\mathbb{B}|b$ (the algebra obtained by intersecting every element in $\mathbb{B}$ with b). A complete Boolean algebra $\mathbb{B}$ is called a <u>Souslin algebra</u> iff $\mathbb{B}$ has a dense subset which is a Souslin tree under the reverse ordering. (More on this concept in Chapter VIII.)

Now it is easy to prove that the canonical complete Boolean algebra constructed from the totally homogeneous Souslin tree of the last paragraph is a homogeneous Souslin algebra with at most ω_1 automorphisms. Using the trees of Theorem 1 and Theorem 3 we obtain in the same way a rigid Souslin algebra and a Souslin algebra with exactly 2 automorphisms. Finally, the definition of the homogeneous Souslin tree with $\geq \omega_2$ automorphisms can be modified so that the tree will be totally homogeneous. Hence we obtain a homogeneous Souslin algebra with at least ω_2 automorphisms.

Chapter VI

MARTIN'S AXIOM AND THE CONSISTENCY OF SH

We have seen earlier that, given a c.t.m. M of ZFC + GCH , there are cardinal absolute generic extensions N_1 and N_2 of M such that

(i) $N_1 \models 2^\omega = \omega_2$ & $\neg$ SH

(ii) $\mathcal{P}^{N_2}(\omega) = \mathcal{P}^M(\omega)$ and $N_2 \models$ GCH & $\neg$ SH .

In this chapter and subsequent ones we shall show how it is possible to obtain models N_1' and N_2' which give the corresponding consistency results for SH in place of $\neg$ SH . The construction of N_1' is well known, and there are at least two good expositions already in the literature, so we shall only sketch the proof, and use this sketch to motivate the construction of N_2' which we shall give in full detail in the remaining chapters. Most of the results we shall mention here are due to Solovay and Tennenbaum. In particular, it was they who first established the relative consistency of SH .

Suppose then we are given a c.t.m. M of ZFC , and we wish to find a cardinal absolute generic extension N of M such that SH holds in M . Thus, in N there will have to be no Souslin trees. In particular, any Souslin tree in M will have to lose its "Souslinity" when we pass to N . Since we will require M and N to have the same cardinals, it is clear that an ω_1-tree in M will still be an ω_1-tree in N , so a Souslin tree $\underset{\sim}{T}$ in M can only lose its "Souslinity" by gaining an uncountable antichain (or, a fortiori, an ω_1-branch) in N . So it is not unreasonable to commence by seeing how we can destroy the Souslinity of single Souslin tree in M . This turns out to be rather easy. However, to simplify the notation (In later chapters, this simplification will be extremely helpful.), let us first of all amend our definition of an ω_1-tree (and hence of an Aronszajn tree and a Souslin tree) so that they "grow downwards" rether than "upwards" as previously. To obtain the amended definition, we simply replace $\leq_T$ by $\geq_T$ everywhere in the original definitions (chapter II). (We shall still talk about "height" however, rather than "depth", and this will mean just what it did before, but in terms of $\geq_T$ rather than $\leq_T$.) The advantage of this amendment is that now a Souslin tree is a splitting poset satisfying c.c.c. Thus, if $\underset{\sim}{T}$ is a Souslin tree in a c.t.m. M of ZFC, then if we force with $\underset{\sim}{T}$, we do not destroy any cardinals. It turns out that we also answer our original question, for we have already proved in Chapter II the following lemma (Theorem II.7)

<u>Lemma 1</u>

Let M be a c.t.m. of ZFC, and let $\underset{\sim}{T}$ be a Souslin tree in M. (Thus, in the real world, $\underset{\sim}{T}$ is an α-tree for $\alpha = \omega_1^M$ a countable limit ordinal, and so $\underset{\sim}{T}$ has 2^ω α-branches.) Let $b \subseteq T$. Then b will be a cofinal branch of $\underset{\sim}{T}$ just in case b is M-generic for $\underset{\sim}{T}$. ❙

Hence, if we take a Souslin tree $\underset{\sim}{T}$ in a c.t.m. M of ZFC and form a generic extension M[G] of M by taking a set G M-generic for $\underset{\sim}{T}$, then $\underset{\sim}{T}$ will not be Souslin in M[G], since G will be, in M[G], a cofinal branch of $\underset{\sim}{T}$. We refer to this process as <u>killing</u> the Souslin tree $\underset{\sim}{T}$. This term is justified because, clearly, if we make <u>any</u> further generic extension N of M[G], then $\underset{\sim}{T}$ will not be a Souslin tree in N either: once a Souslin tree is killed, it is dead and everafter remains dead! This last observation means that a hopeful method for obtaining a model of SH would be to iterate, in some manner, lemma 1, killing all the Souslin trees in sight, one at a time. Now, in order for this to work, it is essential that the entire iteration can be "handled", to some extent, within the original model M : we want our final model to be a model of set theory, and the only way we know how to do this is by the method of forcing. Well, in the case of two generic extensions (and hence any finite number of such) formed successively, this is easily arranged. We describe this arrangement below. However, it turns out to be much easier to consider not the posets, $\underset{\sim}{T}$, concerned, but rather their associated (complete) boolean algebras. (Since forcing with $\underset{\sim}{T}$ is just a disguised form of forcing with $BA(\underset{\sim}{T})$ - see chapter I - this is clearly irrelevant as far as killing $\underset{\sim}{T}$ is concerned.) This will not be apparent in the case of two extensions described in the next lemma, but already for doing ω "successive extensions" the advantage will be quite significant. From now on, we shall use "BA" to denote "complete boolean algebra". The following lemma is a significant strengthening of lemma I.8.

Lemma 2 (Solovay)

Let M be a c.t.m. of ZFC . Let $\mathbb{B}_o$ be a BA in M . Let $\mathbb{B}_1 \in M^{(\mathbb{B}_o)}$ be such that (in M) $\| \mathbb{B}_1 \text{ is a BA} \|^{\mathbb{B}_o} = 1\!\!1$. Then there is, in M , a unique BA $\mathbb{B}_2$ (with $\text{dom}(\mathbb{B}_2) = \{x \in M^{(\mathbb{B}_o)} \mid \| x \in \mathbb{B}_1 \|^{\mathbb{B}_o} = 1\!\!1\}$) such that $M^{(\mathbb{B}_o)(\mathbb{B}_1)} \cong M^{(\mathbb{B}_2)}$, and a canonical complete embedding $e : \mathbb{B}_o \to \mathbb{B}_2$.

Furthermore, whenever $\mathbb{B}_o, \mathbb{B}_1, \mathbb{B}_2$ are related as above, we have:

(i) if G_o is M-generic for $\mathbb{B}_o$ and G_1 is $M[G_o]$-generic for $\mathbb{B}_1/G_o$ (the BA in $M[G_o]$ whose name in the $\mathbb{B}_o$-forcing language is $\mathbb{B}_1$) , then there is a set G_2 which is M-generic for $\mathbb{B}_2$ such that $M[G_o][G_1] = M[G_2]$.

(ii) if G_2 is M-generic for $\mathbb{B}_2$ then (assuming for simplicity that $e = id \upharpoonright \mathbb{B}_o$ in the above) $G_o = G_2 \cap \mathbb{B}_o$ is M-generic for $\mathbb{B}_o$ and there is an $M[G_o]$-generic set G_1 for $\mathbb{B}_1/G_o$ such that $M[G_o][G_1] = M[G_2]$.

Proof: This is just lemma 85 of [Je1]. We shall simply describe the construction of $\mathbb{B}_2$ and e. Let $\mathrm{dom}(\mathbb{B}_2) = \{x \in M^{(\mathbb{B}_0)} \mid \|x \in \mathbb{B}_1\|^{\mathbb{B}_0} = \mathbb{1}\}$, and, for example, to define $x \vee y$ in the sense of $\mathbb{B}_2$, use the maximum principle for $M^{(\mathbb{B}_0)}$ to find a unique (by normality) $z \in M^{(\mathbb{B}_0)}$ such that $z \in \mathrm{dom}(\mathbb{B}_2)$ and $\|z = x \vee y$ (in the sense of $\mathbb{B}_1)\|^{\mathbb{B}_0} = \mathbb{1}$.
Given $b_0 \in \mathbb{B}_0$, let $e(b_0)$ be that unique element $b_2 \in \mathbb{B}_2$ such that $\|b_2 = \mathbb{1}\|^{\mathbb{B}_0} = b_0$ and $\|b_2 = \mathbb{0}\|^{\mathbb{B}_0} = -b_0$.
The rest of the proof is routine, but not trivial. ∎

Lemma 2 clearly enables us to handle any finite number of Souslin trees in one go. That we do not destroy any cardinals in the process is obvious, but for this to remain so when we come to deal with infinitely many Souslin trees, we require the following lemma:

<u>Lemma 3</u> (Solovay)
Let $M, \mathbb{B}_0, \mathbb{B}_1, \mathbb{B}_2$ be related as in lemma 2. If $M \models$ "$\mathbb{B}_0$ has c.c.c." and (in M) $\|$"$\mathbb{B}_1$ has the c.c.c."$\|^{\mathbb{B}_0} = \mathbb{1}$, then $M \models$ "$\mathbb{B}_2$ has the c.c.c."

Proof: The obvious idea works. The details are in [Je1], Lemma 86. ∎

Suppose now we use lemma 2 to deal successively with an ω-sequence of Souslin trees in M. Thus, in M we construct an ω-sequence $\langle \mathbb{B}_n \mid n < \omega \rangle$ of BA's and a commutative system $\langle e_{nm} \mid n < m < \omega \rangle$ of complete embeddings $e_{nm} : \mathbb{B}_n \to \mathbb{B}_m$ (where, for $m > n+1$, e_{nm} is defined by composition). A natural BA (in M) associated with this system is its <u>direct union</u>. This is best described as follows, Isomorph the system so that each e_{nm} is the identity function on $\mathbb{B}_n$. Then $\mathbb{B} = \bigcup_{n<\omega} \mathbb{B}_n$ is a boolean algebra of course. Let $\mathbb{B}_\omega$ be its completion. (One way to form $\mathbb{B}_\omega$ is to regard $\mathbb{B}$ as a poset and set $\mathbb{B}_\omega = \mathrm{BA}(\mathbb{B})$). We call $\mathbb{B}_\omega$ the <u>direct union</u> of the chain

$\langle \mathbb{B}_n \mid n < \omega \rangle$. In the case where the embeddings e_{nm} were not the identity, we call the BA $\mathbb{B}_\omega$, together with the corresponding complete embeddings $e_{n\omega} : \mathbb{B}_n \to \mathbb{B}_\omega$, the direct limit of the system $\langle (\mathbb{B}_n)_{n<\omega}, (e_{nm})_{n<m<\omega} \rangle$. Suppose then we take (in M) $\mathbb{B}_\omega$ as defined here as the limit algebra in our iteration. If G is M-generic for $\mathbb{B}_\omega$, will all of our previous (murderous) work be incorporated in M[G] ? The positive answer to this question is supplied by our next lemma, in conjunction with the second part of lemma 2.

Lemma 4 (Solovay)
Let $\mathbb{B}_o, \mathbb{B}_2$ be BA's in M , and let $e : \mathbb{B}_o \to \mathbb{B}_2$ be a complete embedding in M . Then there is an element $\mathbb{B}_1 \in M^{(\mathbb{B}_o)}$ such that (in M) $\|\mathbb{B}_1 \text{ is a BA}\|^{\mathbb{B}_o} = \mathbb{1}$ and $M^{(\mathbb{B}_o)(\mathbb{B}_1)} \cong M^{(\mathbb{B}_2)}$.

Proof: See [SoTe] for details. Roughly speaking, $\mathbb{B}_1$ is obtained by factoring $\mathbb{B}_2$ through the canonical name (in M) for M-generic ultrafilters on $\mathbb{B}_o$. ∎

That $\mathbb{B}_\omega$ does not destroy any cardinals is guaranteed by the next lemma. Since the above limit construction will work equally well for systems of any (limit) length, we state it in a fairly general way.

Lemma 5 (Solovay - Tennenbaum)
Let α be any limit ordinal, and let $\langle \mathbb{B}_\gamma \mid \gamma < \alpha \rangle$ be an increasing chain of BA's such that (a) $\gamma < \delta < \alpha \to \mathbb{B}_\gamma$ is a complete subalgebra of $\mathbb{B}_\delta$, (b) $\gamma < \alpha \to \mathbb{B}_\gamma$ has c.c.c., and (c) $\gamma < \alpha$ & $\lim(\gamma) \to \mathbb{B}_\gamma$ is the direct union of $\langle \mathbb{B}_\delta \mid \delta < \gamma \rangle$. Let $\mathbb{B}_\alpha$ be the direct union of the $\mathbb{B}_\gamma$'s . Then $\mathbb{B}_\alpha$ has c.c.c.

Proof: By a combinatorial argument. (The only non-trivial case is when $cf(\alpha) = \omega_1$, of course.) The details are given in [Je1], Theorem 49. ∎

Using the above lemmas, it is clear how one could, assuming M is a c.t.m. of ZFC + GCH , obtain a cardinal absolute generic extension N of M in which all of the (at most ω_2 many) Souslin trees in M are killed. To do this one simply commences with an enumeration (in M) of all of the Souslin trees in M , and proceeds to go through them all, killing then off sucessively. (Note that since we are doing everything in M , we don't actually carry out the murder at each stage, we simply arrange it so that the Souslin tree concerned will be included in some final mass-murder, instigated by adding a generic subset of the final BA .) Of course, this will not imply that $N \models$ SH . For at each stage, although we arrange for some Souslin tree to be killed, we may well introduce new Souslin trees in the process. However, it is possible (and essentially just a matter of "bookkeeping" to arrange) to continually change the strategy so as to include any new Souslin trees which may appear. (cf. the "usual proof" that $\omega \times \omega$ is equinumerous with ω .) This is described fully in [Je1] under the title "Model IX".

Now, readers who have been following our reference to [Je1] will have observed that he does not refer to SH , but rather to a (stronger) principle known as Martin's Axiom. In order to motivate the formulation of Martin's Axiom somewhat, we prove the following theorem, which tells us that in any iteration we perform in order to obtain SH in some model, we are (ostensibly) doing something more than just killing Souslin trees, and in face need not consider trees at all, but rather a certain class of posets.

<u>Theorem 6</u>

Let M be a c.t.m. of ZFC . The following are equivalent:

(i) $M \models \neg$ SH .

(ii) There is a (splitting) poset $\mathbb{P}$ in M of cardinality ω_1 such that:

(a) $M \models$ "$\mathbb{P}$ satisfies c.c.c." ;

(b) if G is M-generic for $\mathbb{P}$, then $\mathcal{P}^{M[G]}(\omega) = \mathcal{P}^{M}(\omega)$;

(c) for some set $X \in M$ of cardinality ω_1 , if G is X-generic for $\mathbb{P}$, then $G \notin M$.

Proof: (i) → (ii). Let $\mathbb{P} = \underset{\sim}{T}$ be a Souslin tree in M , and set $X = \{\bigcup_{\beta \geq \alpha} T_\beta \mid \alpha < \omega_1\}$.

(ii) → (i). We give a forcing proof. In chapter VIII we shall, in effect, give an alternative, boolean algebraic proof. Let $\mathbb{P}$ be as in (ii). We may assume $\mathbb{P}$ has domain ω_1 . If $G \subseteq \omega_1^M$ is M-generic for $\mathbb{P}$, we let $F_G : \omega_1^M \to 2$ be the characteristic function for G , with $F_G(\alpha) = 1 \leftrightarrow \alpha \in G$. (Thus $F_G \in M[G]$ for any such G .) In the real world, set $T = \{F_G \restriction \alpha \mid G$ is M-generic for $\mathbb{P}$ & $\alpha < \omega_1^M\}$, and partially order T by $g_1 \leq_T g_2$ iff $g_1 \supseteq g_2$. Thus $\underset{\sim}{T} = \langle T, \leq_T \rangle$ is clearly a tree (in the real world) of height ω_1^M .

<u>Claim 1</u> $\underset{\sim}{T} \in M$.

In M , define $T' = \{f \mid (\exists p \in \mathbb{P})(\exists \alpha < \omega_1)[p \Vdash "\check{f} = F_{\mathring{G}} \restriction \check{\alpha}"]\}$. Suppose $f \in T$. Then for some M-generic G on $\mathbb{P}$ and some $\alpha < \omega_1^M$, $f = F_G \restriction \alpha$, Hence $f \in M[G]$. But $M[G] \models$ "$f \in 2^\alpha \wedge \alpha < \omega_1$" , so we can code f as a subset of ω in $M[G]$. Hence $f \in M$. But look, since $f = F_G \restriction \alpha$, there is $p \in G$ such that $p \Vdash "\check{f} = F_{\mathring{G}} \restriction \check{\alpha}"$. Hence $f \in T'$. Conversely suppose $f \in T'$. Pick $p \in \mathbb{P}$ and $\alpha < \omega_1^M$ so that $p \Vdash "\check{f} = F_{\mathring{G}} \restriction \check{\alpha}"$. Let G be M-generic on $\mathbb{P}$ with $p \in G$. Then $M[G] \models$ "$f = F_G \restriction \alpha$" , so $f \in T$. Hence $T = T' \in M$, and the claim follows immediately (since $\leq_T \, = \, \supseteq$) .

<u>Claim 2</u> $M \models$ "Every antichain of $\underset{\sim}{T}$ is countable".

Let $X \in M$, $X \subseteq T'$, $|X|^M = \omega_1^M$. In M , for each $g \in X$, choose $p(g) \in \mathbb{P}$ and $\alpha(g) < \omega_1$ such that $p(g) \Vdash$ "$\check{g} = F_{\mathring{G}} \restriction \check{\alpha}$". Since $\mathbb{P}$ has c.c.c. in M we can find $g_1, g_2 \in X$ such that $g_1 \neq g_2$ and there is $p \leq p(g_1), p(g_2)$. Thus $p \Vdash$ "$\check{g}_1 = F_{\mathring{G}} \restriction \widetilde{\alpha(g_1)} \wedge \check{g}_2 = F_{\mathring{G}} \restriction \widetilde{\alpha(g_2)}$". Pick G M-generic for $\mathbb{P}$ with $p \in G$. Then, $g_1 = F_G \restriction \alpha(g_1) \wedge g_2 = F_G \restriction \alpha(g_2)$, so either $g_1 \subseteq g_2$ or $g_2 \subseteq g_1$. Hence $M \models$ "X is not an antichain of $\underset{\sim}{T}$", and claim 2 is proved.

We complete the proof by showing that $M \models$ "$\underset{\sim}{T}$ embeds a Souslin tree". In view of the above it clearly suffices to show that $\underset{\sim}{T}$ is a splitting poset. Suppose not. Then, clearly, if $g \in T$ does not split, g lies on a unique cofinal branch b of $\underset{\sim}{T}$. But then there can only be one M-generic subset of $\mathbb{P}$ whose characterictic function extends g, which clearly contradicts the fact that $\mathbb{P}$ is splitting. ▌

By the above, if we are to obtain a model M of SH, then in M it will be the case that: whenever $\mathbb{P}$ is a poset of cardinality ω_1 which satisfies c.c.c., and for which $\Vdash$ "$\mathcal{P}(\check{\omega}) = \widetilde{\mathcal{P}(\omega)}$", and whenever X is a set of cardinality ω_1, then there is (i.e. <u>in M</u>) a set G which is X-generic for $\mathbb{P}$.

Now, if you look back to our sketch of the construction of a model of ZFC + SH, you will observe that nowhere did we make use of the fact that forcing with a Souslin tree adds no new subsets of ω. This leads us, in view of the above, to formulate the following proposition for consideration:

> $\circledast$: Whenever $\mathbb{P}$ is a poset of cardinality ω_1 which satisfies c.c.c., and whenever X is a set of cardinality ω_1, then there is a set G X-generic for $\mathbb{P}$.

Clearly $\circledast \rightarrow$ SH. And it should be clear now that we can, by means of an iterated forcing argument of the type outlined above, obtain a model in which Θ is true. A further analysis of the iteration which makes either SH or Θ true will show that although we may not add new reals at some/all of the sucessor stages, we most certainly will add new reals at limit stages. Hence, we cannot avoid having $2^{\omega} = \omega_2$ in our final model. (In the case of Θ, it is easily seen that in ZFC we can prove outright that $\Theta \to 2^{\omega} \geq \omega_2$: this is just Cohen forcing!) Further consideration along these lines lead to the realisation that there is nothing special about the cardinal ω_1 in Θ, only that it be smaller than the continuum. We thus replace Θ by:

Φ : Whenever $\mathbb{P}$ is a poset of cardinality less than 2^{ω}, which satisfies c.c.c., and whenever X is a set of cardinality less than 2^{ω}, then there is a set G X-generic for $\mathbb{P}$.

It is shown in [Je1] that CH $\to \Phi$. (This is just the usual construction of "X-generic sets" for countable X.) By iterated forcing arguments of the type outlined above, one can establish the consistency of $\Phi + 2^{\omega} = \omega_2$, $\Phi + 2^{\omega} = \omega_3$, etc. (And clearly, $\Phi + 2^{\omega} = \omega_2$ is just Θ.)

Finally, we obtain <u>Martin's Axiom</u>, MA, by omitting from Φ the requirement that $\mathbb{P}$ have cardinality less than 2^{ω}. We may do this because, by a simple application of the downward Løwenheim-Skolem theorem, one sees that $\Phi \to$ MA. (For a "direct" proof, see [Je1], lemma 89.)

MA is an extremely powerful assumption as far as the continuum is concerned, which is of importance in view of the consistency of MA relative to ZFC. In [MaSo], several consequences of MA are obtained. In particular, it is shown that if MA holds, then the union of $< 2^{\omega}$ Lebesgue measure zero sets has Lebesgue measure zero, the union of $< 2^{\omega}$ Lebesgue measurable sets is Lebesgue measurable, and the union of $< 2^{\omega}$ meager sets is meager. (All of these are trivial if $2^{\omega} = \omega_1$, of course, by definition , but they become highly significant when $2^{\omega} > \omega_1$.) As an example of an application of MA we shall show that $MA + 2^{\omega} > \omega_1$ implies that every Aronszajn tree is special. This result is due variously to Kunen, Baumgartner, and Reinhardt-Malitz. It will, in view of our above discussions, serve to establish, in a fairly strong manner, the consistency of $SH + 2^{\omega} = \omega_2$ relative to ZFC . (We shall prove the consistency of $SH + 2^{\omega} = \omega_1$, in the remaining chapters.)

First we require some simple combinatorial facts concerning Aronszajn trees in general.

Let X be an infinite set, n a positive integer. We say $Y \subseteq X^n$ is well-distributed iff, for every finite set $u \subseteq X$ there is $\langle \vec{x} \rangle \in Y$ such that $U \cap \{\vec{x}\} = \emptyset$. Clearly, this condition is equivalent to the existence of an infinite set $Y' \subseteq Y$ such that whenever $\langle \vec{x} \rangle$, $\langle \vec{y} \rangle$ are distinct members of Y' , then $\{\vec{x}\} \cap \{\vec{y}\} = \emptyset$. Given a positive integer m , we shall thus say that $Y \subseteq X^n$ is m-distributed iff there are $\langle \vec{x}_1 \rangle, \ldots, \langle \vec{x}_m \rangle \in Y$ such that $1 \leq i < j \leq m \rightarrow \{\vec{x}_i\} \cap \{\vec{x}_j\} = \emptyset$. Clearly, $Y \subseteq X^n$ will be well-distributed iff it is m-distributed for all $m < \omega$.

If $\underset{\sim}{T}$ is a tree and $z \in S \subseteq T$, $S^{(z)}$ will denote the tree $\{y \in S \mid y \leq_T z \text{ or } z \leq_T y\}$.

Let $\underset{\sim}{T}^1$, $\underset{\sim}{T}^2$ be ω_1-trees. $\underset{\sim}{T}^1 \otimes \underset{\sim}{T}^2$ will denote the ω_1-tree with domain $\{\langle x,y \rangle \mid (\exists \alpha < \omega_1)[x \in T^1_\alpha \ \& \ y \in T^2_\alpha]\}$ and ordering $\langle x,y \rangle \leq \langle x',y' \rangle \leftrightarrow x \leq_1 x' \ \& \ y \leq_2 y'$. (It is clear that $\underset{\sim}{T}^1 \otimes \underset{\sim}{T}^2$ will in fact be an ω_1-tree .) Similarly $\underset{\sim}{T}^1 \otimes \ldots \otimes \underset{\sim}{T}^n$ for any positive integer n . In case $\underset{\sim}{T}^1 = \ldots = \underset{\sim}{T}^n = \underset{\sim}{T}$ here, we write simply $\underset{\sim}{T}^n$ for $\underset{\sim}{T}^1 \otimes \ldots \otimes \underset{\sim}{T}^n$.

The following simple lemma is the crux of our proof :

Lemma 7

Let $\underset{\sim}{T}$ be an Aronszajn tree, n a positive integer. Let S be a final section of $\underset{\sim}{T}^n$ such that every $z \in S$ has uncountably many extensions in S . For every $z \in S$ there is an $\alpha < \omega_1$ such that $(\forall\beta \geq \alpha)(S_\beta^{(z)}$ is well-distributed).

Proof: We shall regard n-tuples as maps with domain n for the purposes of this proof.

Claim 1 Let $z \in S$, $i < n$. For any $p < \omega$ there are $z_1,\dots,z_p \in S$, $z_1,\dots,z_p \leq z$, such that $z_1,\dots,z_p$ are on the same level of S and $1 \leq j < k \leq p \rightarrow z_j(i) \neq z_k(i)$.

Suppose not, and let p be maximal with the above property. Pick $z_1,\dots,z_p$ as stated. Then if $w,w' \leq z_1$, $w,w' \in S$, and $ht(w) = ht(w')$, we must have $w(i) = w'(i)$. Hence $\{w(i) \mid w \in S^{(z_1)}\}$ is an ω_1-branch of $\underset{\sim}{T}$, which is absurd.

Claim 2 Let $z \in S$. There are $z_o,z_1 \in S$, $z_o,z_1 \leq z$, such that $ht(z_o) = ht(z_1)$ and $(z_o"n) \cap (z_1"n) = \emptyset$.

We construct, by induction on $k \leq n$, elements z_o^k,z_1^k of S such that $z_o^k,z_1^k \leq z$, $ht(z_o^k) = ht(z_1^k)$, and $(z_o^k"k) \cap (z_1^k"k) = \emptyset$, whence $z_o = z_o^n$ and $z_1 = z_1^n$ are as required.
Set $z_o^o = z_1^o = z$. Suppose $k < n$ and z_o^k, z_1^k, are defined. By claim 1, pick $w_o,\dots,w_n \in S$, $w_o,\dots,w_n \leq z_o^k$, $ht(w_o) = \dots = ht(w_n)$, such that $0 \leq i < j \leq n \rightarrow w_i(k) \neq w_j(k)$. Pick $v \in S$, $v \leq z_1^k$, $ht(v) = ht(w_o)$. For all $i \leq n$, $(v"k) \cap (w_i"k) = \emptyset$ of course, as $\underset{\sim}{T}$ is a tree . By cardinality considerations, there must be $i \leq n$ such that $w_i(k) \notin v"n$. Let i be the least such. Set $z_o^{k+1} = w_i$, $z_1^{k+1} = v$.

Claim 3 Let $z \in S$. For each $m < \omega$ there is $\alpha_m^z < \omega_1$ such that $z \in S|\alpha_m^z$ and $S_{\alpha_m^z}^{(z)}$ is m-distributed.

This follows from claim 2 by a simple induction on m .

Suppose now that $z \in S$ is given. Set $\alpha^z = \sup_{m<\omega} \alpha_m^z$. Clearly, if $\beta \geq \alpha^z$, then $S_\beta^{(z)}$ is m-distributed for all m, and hence well-distributed. ▌

Corollary 8

Let $\underset{\sim}{T}$, n, S be as above. There is a strictly increasing, continuour function $\tau : \omega_1 \to \omega_1$ such that whenever $z \in S_{\tau(\nu)}$, then $(\forall \eta > \nu)(S_{\tau(\eta)}^{(z)}$ is well-distributed).

Proof: Immediate. ▌

Theorem 9

Assume $MA + 2^\omega > \omega_1$. Then every Aronszajn tree is special.

Proof: Let $\underset{\sim}{T}$ be any Aronszajn tree. We say u is a neat subtree of $\underset{\sim}{T}$ iff u is a finite subtree of $\underset{\sim}{T}$ such that:

(i) if $x, y \in u$ have the same height in $\underset{\sim}{T}$, then they have the same height in u, and conversely; and

(ii) every point of u has an extension on every level of u.

A partial embedding (p.e.) of $\underset{\sim}{T}$ into $\mathbb{Q}$ is an order-preserving map p from a neat subtree of $\underset{\sim}{T}$ into $\mathbb{Q}$. Let $\mathbb{P}$ be the poset of all p.e.'s ordered by inclusion.

Claim $\mathbb{P}$ satisfies c.c.c.

Ignoring the proof of this claim for the time being, let us observe how the theorem follows from it. For each $x \in T$, let $D_x = \{p \in \mathbb{P} \mid x \in \text{dom}(p)\}$. Clearly D_x is dense in $\mathbb{P}$. Let $X = \{D_x \mid x \in T\}$. Since $|X| = \omega_1 < 2^\omega$, by MA there is an X-generic set G for $\mathbb{P}$. Clearly $f = \bigcup G$ is an order-preserving map from $\underset{\sim}{T}$ into $\mathbb{Q}$, so $\underset{\sim}{T}$ is special.

We turn now to the proof of the claim. Let $X \subseteq \mathbb{P}$ be uncountable. We show that X must contain a pair of compatible elements. Firstly, we can clearly assume that the domains of all of the members of X are isomorphic (as neat subtrees of $\underset{\sim}{T}$). Secondly, we can (equally clearly) assume that if $p,q \in X$ and $x \in \mathrm{dom}(p)$, $y \in \mathrm{dom}(q)$ correspond to one another under the isomorphism of $\mathrm{dom}(p)$ and $\mathrm{dom}(g)$, then $p(x) = q(y)$. (Of course, there may be several such isomorphisms for each such pair p,q, but we can eliminate this ambiguity by, for example, linearly ordering each level of $\underset{\sim}{T}$ in some arbitrary, but coherent manner, and taking just those isomorphisms which preserve these orderings.) Let n be the number of levels in the domain of any member of X. Pick $k \leq n$ largest so that uncountably many members of X have a domain which end-extends a single neat subtree with k levels. If $k = n$ we are done, so assume $k < n$ and that u_0 is a neat subtree of $\underset{\sim}{T}$ with k levels such that every member of X has domain an end-extension of u_0. We may assume, by choice of k, that the $(k+1)$'st level of the domain of each member of X lies on a $\underset{\sim}{T}$-level unique for that member. Let m be the cardinality of the $(k+1)$'st level of the domain of each member of X.

Let U be the set of all $(k+1)$'st levels of all $\mathrm{dom}(p)$ for $p \in X$. U determines, in a natural way (eg. using our linear orderings of the levels of $\underset{\sim}{T}$), a final section S of $\underset{\sim}{T}^m$. Since each level of $\underset{\sim}{T}^m$ is countable, we can assume that every member of S has uncountably many extensions in S. (Throw away those members of X which do not provide such nice points!) Pick an arbitrary member $\langle\vec{z}\rangle$ of S such that $\{\vec{z}\} \in U$. Let $\alpha = ht_S(\langle\vec{z}\rangle)$. By lemma 7, we can find $\beta > \alpha$ such that $S_\beta^{\langle\vec{z}\rangle}$ is well-distributed. Pick extensions $\langle\vec{z}_1\rangle$,

$\langle \vec{z}_2 \rangle$ of $\langle \vec{z} \rangle$ in S_β so that $\{\vec{z}_1\} \cap \{\vec{z}_2\} = \emptyset$. Let $\{\vec{y}_1\}$, $\{\vec{y}_2\} \in U$, be such that $\langle \vec{y}_1 \rangle$ extends $\langle \vec{z}_1 \rangle$ in S and $\langle \vec{y}_2 \rangle$ extends $\langle \vec{z}_2 \rangle$ in S.

Let p_1, p_2 be those members of X such that $\{\vec{y}_1\}$, $\{\vec{y}_2\}$ are their respective (k+1)'st levels. Clearly, $\mathrm{dom}(p_1) \cap \mathrm{dom}(p_2) = u_0$, and if $x \in \mathrm{dom}(p_1)$, $y \in \mathrm{dom}(p_2)$, $x,y \notin u_0$, then x and y are incomparable in $\underset{\sim}{T}$. Hence $p_1 \cup p_2$ extends to an $r \in \mathbb{P}$, and we are done. ▌

<u>Remark</u> The proofs of lemma 7 and theorem 9 seemed quite long bec use we gave them in some detail. In essence, both were extremely simple. Since the ideas involved will be used again in a much more complicated situation, the reader is advised to gain a clear picture of what is going on here before proceeding any further. (We might also remark that the proofs of theorem 9 which Kunen, Baumgartner, and Reinhardt-Malitz gave were somewhat more direct than ours, which we developed with an eye to the immediate future!)

We end this chapter with a remark concerning the joint consistency of SH + CH. Suppose we obtain a model, M, of $SH + 2^{\omega} = \omega_2$ as above. If we form a boolean extension N of M such that $N \models CH$ by collapsing ω_2^M to ω_1^M in the usual way, then by an argument as in theorem III.6, we see at once that $N \models \Diamond$, so that SH will fail in N. Hence, if we are to obtain a model of SH + CH, we must adopt a different approach. We shall commence to do this in the next chapter.

Chapter VII

TOWARDS CON(ZFC + CH + SH) : A FALSE START

We begin this chapter by describing a particular situation where it is possible to carry out an iterated forcing argument without introducing new reals.

Let $\mathbb{P}$ be a poset, κ a cardinal. $\mathbb{P}$ is said to be neatly κ-closed if whenever $\langle p_\alpha \mid \alpha < \beta < \kappa \rangle$ is a decreasing sequence from $\mathbb{P}$, then $p \in \mathbb{P}$, where $p = \bigwedge_{\alpha<\beta} p_\alpha$ in $BA(\mathbb{P})$.

Clearly, any κ-closed normal tree will be neatly κ-closed, and the same is true of many (but not all) of the posets used in forcing. Also, of course, if $\mathbb{P}$ is neatly κ-closed, then $\mathbb{P}$ is certainly κ-closed in the sense of chapter I.

We shall require a slight sharpening of Lemma I.7. Note that a poset $\mathbb{P}$ will be κ-dense just in case $BA(\mathbb{P})$ is a κ-dense poset. Recall that a BA $\mathbb{B}$ is (κ,∞)-distributive iff for all $\beta < \kappa$ and all $\gamma \in On$, if $\{b_{\nu\tau} \mid \nu < \beta, \tau < \gamma\} \subseteq \mathbb{B}$, then $\bigwedge_{\nu<\beta} \bigvee_{\tau<\gamma} b_{\nu\tau} = \bigvee_{f\in\gamma^\beta} \bigwedge_{\nu<\beta} b_{\nu,f(\nu)}$.

Lemma 1

Let κ be an uncountable regular cardinal $\mathbb{B}$ a BA . The following are equivalent:

(i) $\mathbb{B}$ is κ-dense;

(ii) $\mathbb{B}$ is (κ,∞)-distributive;

(iii) for all $\alpha < \kappa$, $\|\check{V}^{\check{\alpha}} = \check{(V^\alpha)}\|^{\mathbb{B}} = \mathbb{1}$.

Proof: (Sketch) (i) $\rightarrow$ (iii). Let $b \leq \|u \in \check{V}^{\check{\alpha}}\|$. For each $\xi < \alpha$, $D_\xi = \{b' \in \mathbb{B} \mid b' \wedge b = \mathbb{0}$ or $b' \leq \|u(\check{\xi}) = \check{x(\xi)}\|$ for some $x(\xi) \in V\}$ is a dense initial section of $\mathbb{B}$. Let $d \leq b$, $d \in \bigcap_{\xi<\alpha} D_\xi$. Then $d \leq \|u = \check{x}\|$ for some $x \in V$, $x : \alpha \rightarrow V$.

(iii) → (ii). Let $\beta < \varkappa$, $\gamma \in On$, $\{b_{\nu\tau} | \nu < \beta, \tau < \gamma\} \subseteq \mathbb{B}$. It is easily seen that $\bigvee_{f\in\gamma^\beta} \bigwedge_{\nu<\beta} b_{\nu,f(\nu)} \leq \bigwedge_{\nu<\beta} \bigvee_{\tau<\gamma} b_{\nu\tau} = b$, say. To prove the opposite inequality, define $g \in V^{(\mathbb{B})}$ so that $\|g : \check{\beta} \to \check{\gamma}\| = b$ and $\|g(\check{\nu}) = \check{\tau}\| = b_{\nu\tau} - \bigvee_{\eta<\tau} b_{\nu\eta}$ for each ν, τ . Then by (iii), $b \leq \|(\exists f \in \check{\gamma^\beta})(\forall \nu \in \check{\beta})(f(\nu) = g(\nu))\| \leq$ $\bigvee_{f\in\gamma^\beta} \bigwedge_{\nu<\beta} b_{\nu,f(\nu)}$.

(ii) → (i). Let D_ν , $\nu < \beta < \varkappa$ be dense initial sections of $\mathbb{B}$. Let $\langle b_{\nu\tau} | \nu < \gamma \rangle$ enumerate (possibly with repetitions) D_ν , each ν . By density, $\bigwedge_{\nu<\beta} \bigvee \{b_{\nu\tau} | \tau < \gamma\} = \mathbb{1}$, so $\bigvee_{f\in\gamma^\beta} \bigwedge_{\nu<\beta} b_{\nu,f(\nu)} = \mathbb{1}$. Hence if $b \in \mathbb{B}$ then $b \wedge \bigwedge_{\nu<\beta} b_{\nu,f(\nu)} = c > \mathbb{0}$ for some $f \in \gamma^\beta$, which implies that $c \leq b$, $c \in \bigcap_{\nu<\beta} D_\nu$. ▮

Let $\mathbb{B}_0, \mathbb{B}_1$ be BA's , $e : \mathbb{B}_0 \to \mathbb{B}_1$ a complete embedding. The e-basic projection $h : \mathbb{B}_1 \to \mathbb{B}_0$ is defined by $h(b_1) = \bigwedge\{b_0 \in \mathbb{B}_0 | b_1 \leq_1 e(b_0)\}$. (If $e = id \restriction \mathbb{B}_0$, we call h the basic projection.)

Let $\mathbb{P}_0, \mathbb{P}_1$ be posets. A map $f : \mathbb{P}_1 \to \mathbb{P}_0$ is a neat cover iff: (i) f is surjective; (ii) $p \leq_1 q \to f(p) \leq_0 f(q)$; (iii) $p \leq_0 f(q) \to (\exists q' \leq_1 q)(p = f(q'))$.

We shall require the following stronger version of lemma VI.2.

Lemma 2 (Solovay)

Let M be a c.t.m. of ZFC , $\varkappa$ an uncountable regular cardinal in M . Let $\mathbb{P}_0$ be, in M , a neatly $\varkappa$-closed poset, $\mathbb{B}_0 = BA(\mathbb{P}_0)$. Let $\mathbb{P}_1, \mathbb{B}_1 \in M^{\mathbb{B}_0}$ be such that, in M , $\|$"$\mathbb{P}_1$ is a neatly $\check{\varkappa}$-closed poset and $\mathbb{B}_1 = BA(\mathbb{P}_1)$"$\|^{\mathbb{B}_0} = \mathbb{1}$. Then there is, in M , a neatly $\varkappa$-closed poset $\mathbb{P}_2$ and a canonical complete embedding $e : \mathbb{B}_0 \to \mathbb{B}_2 = BA(\mathbb{P}_2)$ such that (in M) :

(i) $e''\mathbb{P}_0 \subseteq \mathbb{P}_2$ and $e(\mathbb{1}_0) = \mathbb{1}_2$;

(ii) if $h : \mathbb{B}_2 \to \mathbb{B}_0$ is the e-basic projection, then $(h \restriction \mathbb{P}_2) : \mathbb{P}_2 \to \mathbb{P}_0$

is a neat cover ;

(iii) if $\{p_\xi \mid \xi < \gamma < \varkappa\} \subseteq \mathbb{P}_2$ **is decreasing** , then $h(\wedge_{\xi<\gamma} p_\xi) = \wedge_{\xi<\gamma} h(p_\xi)$;

[We call any system $(\mathbb{P}_0, \mathbb{B}_0; \mathbb{P}_2, \mathbb{B}_2; e, h)$ so related a <u>$\varkappa$-happy system</u>] And if G_0 is M-generic for $\mathbb{P}_0$ and G_1 is $M[G_0]$-generic for $\mathbb{P}_1$ ($\in M[G_0]$), there is G_2 , which is M-generic for $\mathbb{P}_2$, such that $M[G_0][G_1] = [G_2]$, and conversely.

Proof: We assume $M^{\mathbb{B}_0}$ is normalized. Thus, in M, $Q = \{p \in M^{\mathbb{B}_0} \mid \|p \in \mathbb{P}_1\|^{\mathbb{B}_0} > \mathbb{0}\}$ is a set. Work in M . Let $\mathbb{P}_2$ have domain $\{\langle p,q\rangle \mid p \in \mathbb{P}_0 \ \& \ q \in Q \ \& \ p \Vdash_0 \text{"} q \in \mathbb{P}_1\text{"}\}$. Set $\langle p',q'\rangle \leq_2 \langle p,q\rangle \leftrightarrow p' \leq_0 p \ \& \ p' \Vdash_0 \text{"} q' \leq_2 q\text{"}$. It is easily seen that $\mathbb{P}_2$ is a neatly $\varkappa$-closed poset. In particular, if $\langle\langle p_\alpha, q_\alpha\rangle \mid \alpha < \lambda < \varkappa\rangle$ is a decreasing sequence from $\mathbb{P}_2$, then $p = \wedge_{\alpha<\lambda} p_\alpha \in \mathbb{P}_0$ (where this inf is taken in $\mathbb{B}_0$ of course), and thus $p \Vdash_0 \text{"}\langle q_\alpha \mid \alpha < \check{\lambda}\rangle$ is a decreasing sequence from $\mathbb{P}_1$", so by the maximum principle for $M^{\mathbb{B}_0}$ there is a unique $q \in Q$ such that $p \Vdash_0 \text{"} q = \wedge_{\alpha<\check{\lambda}} q_\alpha$ (in $\mathbb{B}_1$)" . Clearly, $\langle p,q\rangle = \wedge_{\alpha<\lambda}\langle p_\alpha, q_\alpha\rangle$ (in $\mathbb{B}_2 = BA(\mathbb{P}_2)$). **This also proves (iii) of course.**

Define $f : \mathbb{P}_2 \to \mathbb{P}_0$ by $f(\langle p,q\rangle) = p$. Clearly, f is a neat cover. Define $e : \mathbb{B}_0 \to \mathbb{B}_2$ by setting, for $p \in \mathbb{P}_0$, $e(p) = \{q \in \mathbb{P}_2 \mid f(q) \leq_0 p\}$ (and extending e to $\mathbb{B}_0$ uniquely by the density of $\mathbb{P}_0$ in $\mathbb{B}_0$). Since f is a neat cover, it is easily seen that e is a complete embedding. And clearly, if h is the e-basic projection, then $h \upharpoonright \mathbb{P}_2 = f$. The rest of the lemma follows easily now. ∎

Suppose now that (for definiteness and simplicity) $\varkappa = \omega_1$ in the above, and that, using lemma 2, we constructed ω-chains $\langle \mathbb{P}_n \mid n < \omega\rangle$, $\langle \mathbb{B}_n \mid n < \omega\rangle$, $\langle e_{nm} \mid n < m < \omega\rangle$, $\langle h_{mn} \mid n < m < \omega\rangle$ such that for each n ,

$(\mathbb{P}_n, \mathbb{B}_n; \mathbb{P}_{n+1}, \mathbb{B}_{n+1}; e_{n,n+1}, h_{n+1,n})$ is an ω_1-happy system. (Then clearly, for each pair $n < m < \omega$, $(\mathbb{P}_n, \mathbb{B}_n; \mathbb{P}_m, \mathbb{B}_m; e_{nm}, h_{mn})$ is also ω_1-happy.) Now, by Lemma 1, we know that if we force with any of the $\mathbb{B}_n$'s, we add no new subsets of ω. Can we define a limit operation, suitable for iterated forcing, which also adds no new subsets of ω? In fact we can do so very simply, utilising the fact that the entire ω-chain of systems is ω_1-happy. We first of all set $\mathbb{P}_\omega = \{\langle p_n | n < \omega \rangle | (\forall n < \omega)(p_n \in \mathbb{P}_n \ \&\ p_n = h_{n+1,n}(p_{n+1}))\}$ and partially order $\mathbb{P}_\omega$ by $p \leq_\omega q \leftrightarrow (\forall n < \omega)(p_n \leq_n q_n)$. It is easily verified that $\mathbb{P}_\omega$ is a neatly ω_1-closed poset. Set $\mathbb{B}_\omega = BA(\mathbb{P}_\omega)$. In order that $\mathbb{B}_\omega$ be a suitable limit operation for iterated forcing, we must ensure that each $\mathbb{B}_n$ is completely embeddable into $\mathbb{B}_\omega$. To see this, first of all define $f_n : \mathbb{P}_\omega \to \mathbb{P}_n$ by $f_n(p) = p_n$. It is easily checked that f_n is a neat cover. (Since $p = \langle h_{no}(p_n), h_{n1}(p_n), \ldots, p_n, p_n, \ldots \rangle \in \mathbb{P}_\omega$ for any given $p_n \in \mathbb{P}_n$, condition (i) of the neat cover definition is satisfied; condition (ii) is immediate by the definition of $\leq_\omega$; and for condition (iii), all that is required is a simple induction argument, using the fact that for each $n < \omega$, the basic projection $h_{n+1,n}$ is a neat cover for $\mathbb{P}_{n+1}, \mathbb{P}_n$.) It follows that if we define $e_{n\omega} : \mathbb{B}_n \to \mathbb{B}_\omega$ by setting $e_{n\omega}(p) = \{q \in \mathbb{P}_\omega | f_n(q) \leq_n p\}$ for each $p \in \mathbb{P}_n$ and then extending $e_{n\omega}$ to $\mathbb{B}_n$ by density, then $e_{n\omega}$ is a complete embedding. Furthermore, if $h_{\omega n}$ denotes the $e_{n\omega}$-basic projection of $\mathbb{B}_\omega$ onto $\mathbb{B}_n$, then we clearly have $h_{\omega n} \upharpoonright \mathbb{P}_\omega = f_n$. Hence, not only is $\mathbb{B}_\omega$ a suitable limit of the $\mathbb{B}_n$'s, but also for each $n < \omega$, $(\mathbb{P}_n, \mathbb{B}_n; \mathbb{P}_\omega, \mathbb{B}_\omega; e_{n\omega}, h_{\omega n})$ is ω_1-happy, so that we can at once continue the process. We call $\mathbb{B}_\omega$, together with the embeddings $e_{n\omega}$, $n < \omega$, the <u>inverse limit</u> of the system $\langle (\mathbb{B}_n)_{n<\omega}, (\mathbb{P}_n)_{n<\omega}, (e_{nm})_{n<m<\omega} \rangle$.

We can then iterate forcing with neatly ω_1-closed posets as follows. At successor stages we use lemma 2. At limit stages of cofinality ω, we take a cofinal ω-sequence of our system so far, and form the inverse limit. At limit stages of cofinality greater than ω take the direct limit. This is easily seen to be neatly ω_1-closed and to preserve happiness. (In particular, if we have $\lim(\beta)$, $cf(\beta) > \omega$, and if $\langle (\mathbb{P}_\alpha, \mathbb{B}_\alpha) \mid \alpha < \beta \rangle$, $\langle (e_{\gamma\delta}, h_{\delta\gamma}) \mid \gamma < \delta < \beta \rangle$ denotes the system constructed so far, then if $\langle \mathbb{B}_\beta, (e_{\gamma\beta})_{\gamma<\beta} \rangle$ is the direct limit, $\mathbb{P}_\beta = \{\Lambda_{\alpha<\beta} e_{\alpha\beta}(p_\alpha) \mid (\forall\alpha < \beta)(p_\alpha \in \mathbb{P}_\alpha)$ & $(\forall\gamma < \delta < \beta)$ $(p_\gamma = h_{\delta\gamma}(p_\delta))$ & $\Lambda_{\alpha<\beta} e_{\alpha\beta}(p_\alpha) > \mathbb{0}\}$ is easily seen to be a neatly ω_1-closed dense subset of $\mathbb{B}_\beta$.)

Very well, then, we now know of a situation where we can do iterated forcing without introducing new reals. Can we apply it to the problem of destroying Souslin trees ? Well, a Souslin tree is certainly not ω_1-closed; the best that we have along these lines is the following lemma.

Lemma 3

If $\underset{\sim}{T}$ is a Souslin tree, then $\underset{\sim}{T}$ is ω_1-dense.

Proof: We observed during the proof of Theorem II.7 that if D is a dense initial section of $\underset{\sim}{T}$ then $\bigcup_{\beta \geq \alpha} T_\beta \subseteq D$ for some $\alpha < \omega_1$. The lemma follows immediately from this fact. ▮

Perhaps we can modify the inverse limit to preserve just ω_1-density? In fact this trail is doomed from the start, for Jensen proved long ago that there is no boolean limit operation, suitable for an iterated forcing construction over L in which Souslin trees are killed by being given (generic) cofinal branches, which does not introduce new subsets of ω. (This is an immediate consequence of the proof of the main result in [JnJo]. A sketch of the proof is given as an Appendix.)

A next attempt might be inspired by the proof of theorem VI.9. Suppose we decide to kill Souslin trees now by (generically) order-embedding them into $\mathbb{Q}$, the rationals (say by means of some poset of countable partial embeddings). Assuming we can find a neatly ω_1-closed poset which does this <u>without</u> destroying cardinals, then perhaps we shall succeed in performing a successful iteration? This hope is dashed at once by the following lemma:

<u>Lemma 4</u>

Let M be a c.t.m. of ZFC , and let $\underset{\sim}{T}$ be a Souslin tree in M . Suppose $\mathbb{P}$ is a poset in M such that $M \models$ "$\mathbb{P}$ is ω_1-closed" . Then, if G is M-generic for $\mathbb{P}$, $\underset{\sim}{T}$ is still a Souslin tree in M[G] .

Proof: Work in M . Suppose $p \Vdash$ "$\mathring{X}$ is a maximal antichain of $\check{\underset{\sim}{T}}$". Let $\langle x_\nu \mid \nu < \omega_1 \rangle$ enumerate T . Pick a decreasing sequence $\langle p_\nu \mid \nu < \omega_1 \rangle$ from $\mathbb{P}$ such that $p_0 \leq p$ and for each $\nu < \omega_1$ there is $X_\nu \subseteq T|\nu$ such that $p_\nu \Vdash$ "$\mathring{X} \cap (T|\nu)^\vee = \check{X}_\nu$" and for some y which is $\underset{\sim}{T}$ - comparable with x_ν , $p \Vdash$ "$\check{y} \in \mathring{X}$" . This is clearly possible, since $\mathbb{P}$ is ω_1-closed.
Let $\bar{X} = \bigcup_{\nu<\omega_1} X_\nu$. Clearly, $\bar{X}$ is an antichain of $\underset{\sim}{T}$, so we must have $\bar{X} = X_\alpha$ for some $\alpha < \omega_1$. But by our construction, $\bar{X}$ is also maximal, so X_α is a maximal antichain of $\underset{\sim}{T}$, which means that $p_\alpha \Vdash$ "$\mathring{X} = \check{X}_\alpha$" . Since $p_\alpha \leq p$ and $p_\alpha \Vdash$ "$\mathring{X}$ is countable" , we are done. ▮

Thoroughly disheartened now, we decide it is best to start another chapter.

Chapter VIII

ITERATED FORCING JENSEN STYLE

In this chapter we shall describe Jensen's method for carrying through an iterated forcing argument (using posets which are not ω_1-closed) without adding new reals or collapsing cardinals. Lemma VII.3 provides the key. We shall carry out the iteration so that at each stage - including limit stages - we are forcing with a Souslin tree. Now, in view of the result of Jensen we mentioned in the last chapter (namely that one cannot get SH to hold by iterating the process of adding generic branches, without adding reals), this might at first sight seem impossible unless we forego our desire not to add new reals. However, there is a subtle difference between the two approaches. Jensen's result tells us that the procedure we adopt in destroying a given Souslin tree must not be to force with that tree, or even to force with anything which will add a cofinal branch to it. There is nothing to prevent us from forcing with a Souslin tree (of necessity not the given one) which causes the given tree to be generically order-embedded into $\mathbb{Q}$, say. Of course, the tree which we do force with will lose its Souslinity in the usual manner by gaining a generic branch, but it will prosses sufficient properties (in relation to the iteration) to avoid the Jensen example. We thus have the rather paradoxical strategy that, in order to destroy all Souslin trees, we have to put all our efforts into constructing Souslin trees. (Just how great these efforts will be, will be seen in the next chapter!)

In order to carry out the iteration, it is necessary to first set up a suitable combinatorial framework. This is done by performing an initial generic extension (which does not add new reals or collapse cardinals) in which the combinatorial principles $\Diamond^*$ and $\Box$ hold.

For further details concerning these principles, we refer the reader to [De 1]. We shall here content ourselves with proving just what we need, and no more. We commence with $\square$.

$\square$: There is a sequence $\langle A_\lambda \mid \lambda<\omega_2 \ \& \ \lim(\lambda)\rangle$ such that for all limit ordinals $\lambda < \omega_2$,

(i) A_λ is a closed unbounded subset of λ ;

(ii) if $cf(\lambda) < \omega_1$, then $otp(A_\lambda) < \omega_1$;

(iii) if $\alpha \in A_\lambda$ and $\alpha = \sup(A_\lambda \cap \alpha)$, then $A_\alpha = A_\lambda \cap \alpha$.
(Hence $cf(\lambda) = \omega_1 \rightarrow otp(A_\lambda) = \omega_1$.)

Lemma 1

Let M be a c.t.m. of $ZFC + (2^\omega=\omega_1) + (2^{\omega_1}=\omega_2)$. There is a generic extension N of M such that

(i) M and N have the same cardinalities and cofinality function;

(ii) for all cardinals λ of M , $(2^\lambda)^N = (2^\lambda)^M$;

(iii) $\mathcal{P}^N(\omega) = \mathcal{P}^M(\omega)$ & $\mathcal{P}^N(\omega_1) = \mathcal{P}^M(\omega_1)$;

(iv) $N \models \square$.

Proof: Work in M : Let P be the set of all sequences p whose domain is a proper, closed, initial segment of the limit ordinals below ω_2 , satisfying (i) - (iii) of $\square$. For $p,q \in P$, say $p \leq q \leftrightarrow p \supseteq q$. Let $\mathbb{P}$ be the poset $\langle P, \leq\rangle$. (Let G be M-generic on $\mathbb{P}$ and set $N = M[G]$.). Then, $|P| = \omega_2$, and it is clear that there are only two non-standard parts of the proof. These are:

(1) If $p \in P$ and $\lambda < \omega_2$, $\lim(\lambda)$, there is $q \leq p$ such that $\lambda \in dom(q)$.

(2) Suppose $\langle D_\nu \mid \nu < \omega_1\rangle$ are dense initial sections of $\mathbb{P}$ and that $p \in \mathbb{P}$. Then there is $q \in \mathbb{P}$, $q \leq p$, such that $q \in \bigcap_{\nu<\omega_1} D_\nu$.

Given (1) and (2), the lemma follows easily.

To prove (1), what we do is prove by induction on the limit ordinals $\nu < \omega_2$ that: "for all limit ordinals $\tau < \nu$ and all $p \in \mathbb{P}$ with $\max(\mathrm{dom}(p)) = \tau$, there is $q \leq p$ such that $\max(\mathrm{dom}(q)) = \nu$". Successor stages in the induction are trivial, since we can just add a disjoint ω-sequence. At limit stages, we pick a cofinal cofinality-sequence and work along it in the same way as described below in verifying (2).

To prove (2), we inductively define a sequence $\langle q_\nu | \nu \leq \omega_1 \rangle$ of members of $\mathbb{P}$ and a normal function $\tau : \omega_1+1 \to \omega_2$. Set $q_0 = p$, $\tau(0) = \max(\mathrm{dom}(q_0))$. If q_ν, $\tau(\nu)$ are defined, $\nu < \omega_1$, let $q_{\nu+1}$ be a proper extension of q_ν such that $q_{\nu+1} \in D_\nu$, and set $\tau(\nu+1) = \max(\mathrm{dom}(q_{\nu+1}))$. If $\lim(\nu)$, $\nu \leq \omega_1$, and $\langle q_\eta | \eta < \nu \rangle$, $\tau \upharpoonright \nu$ are defined, set $\tau(\nu) = \sup_{\eta<\nu} \tau(\eta)$ and $q_\nu = (\bigcup_{\eta<\nu} q_\eta) \cup \{\langle \tau''\nu, \tau(\nu) \rangle\}$. It is easily seen that $q = q_{\omega_1}$ is as required. ▮

Our next result ensures that $\Box$ will remain valid when we force to obtain $\Diamond^*$.

<u>Lemma 2</u>

Let M be a c.t.m. of ZFC + $\Box$. If N is any generic extension of M with the same cardinalities and cofinality function as M, then $\Box$ holds in N also.

Proof: Trivial. ▮

$\Diamond^*$: There is a sequence $\langle W_\nu | \nu < \omega_1 \rangle$ such that each W_ν is a countable subset of $\mathcal{P}(\nu)$, and for any $A \subseteq \omega_1$, $\{\alpha \in \omega_1 | A \cap \alpha \in W_\alpha\}$ contains a closed unbounded set.

Lemma 3

Let M be a c.t.m. of ZFC + $(2^{\omega} = \omega_1)$ + $(2^{\omega_1} = \omega_2)$. There is a generic extension N of M such that:

(i) M and N have the same cardinalities and cofinality function;

(ii) for all cardinals λ of M , $(2^{\lambda})^N = (2^{\lambda})^M$;

(iii) $\mathcal{P}^N(\omega) = \mathcal{P}^M(\omega)$;

(iv) $N \models \Diamond^*$.

Proof: Work in M at first: Let P be the set of all pairs $\langle w,B\rangle$ such that w is a map with domain some $\xi(w) < \omega_1$, and for all $\nu < \xi(w)$, $w(\nu)$ is a countable subset of $\mathcal{P}(\nu)$, and such that B is a countable subset of $\mathcal{P}(\omega_1)$. For $\langle w,B\rangle$, $\langle w',B'\rangle$ in P , say $\langle w',B'\rangle \leq \langle w,B\rangle$ iff $w' \supseteq w$ & $B' \supseteq B$ & $(\forall \nu \in \xi(w') - \xi(w))(\forall b \in B)(b \cap \nu \in w'(\nu))$. Let $\mathbb{P} = \langle P,\leq\rangle$. (Let G be M-generic on $\mathbb{P}$ and set $N = M[G]$.) Then $|\mathbb{P}| = \omega_2$ and $\mathbb{P}$ satisfies the ω_2 chain condition, this last because $\langle w,B\rangle$ incompatible with $\langle w',B'\rangle$ implies $w \neq w'$. Hence, as $\mathbb{P}$ is clearly ω_1-closed also, the only non-standard part of the proof is the verification that $\Diamond^*$ will hold in N .

Work in N : Set $W = \bigcup\{w \mid \langle w,\emptyset\rangle \in G\}$. Clearly, $G = \{\langle w,B\rangle \mid w \subseteq W$ & $(\forall \nu \geq \xi(w))(\forall b \in B)(b \cap \nu \in W(\nu))\}$, so $N = M[W]$.

For each $\alpha < \omega_1$, let $\delta(\alpha) < \omega_1$ be the least δ such that for all $b \in W(\alpha)$:

(i) if α is countable in $L[b,W\restriction\alpha]$, then α is countable in $L_{\delta}[b,W\restriction\alpha]$;

(ii) if α is uncountable in $L[b,W\restriction\alpha]$ but $(\alpha^+)^{L[b,W\restriction\alpha]} < \omega_1$, then $\delta \geq (\alpha^+)^{L[b,W\restriction\alpha]}$.

For each $\alpha < \omega_1$, set $\bar{W}_\alpha = \mathcal{P}(\alpha) \cap (\bigcup_{b \in W(\alpha)} L_{\delta(\alpha)}[b, W\restriction\alpha])$, a countable subset of $\mathcal{P}(\alpha)$. We show that $\langle \bar{W}_\alpha | \alpha < \omega_1 \rangle$ realises $\Diamond^*$.

Let $A \subseteq \omega_1$ be given. We can easily find a set $X \subseteq \omega_1$, $X \in M$, such that $A \in L[X,W]$ and $\omega_1 = \omega_1^{L[X,W]}$. Let θ be the least ordinal below ω_1 such that $\omega_1 = \omega_1^{L[X\cap\theta, W\restriction\theta]}$ if such exists, and set $\theta = \omega$ otherwise. Let $\gamma = \omega_2^{L[X,W]}$. Define, inductively, submodels $N_\nu \prec L_\gamma[X,W]$, $\nu < \omega_1$, as follows.

N_0 = the smallest $N \prec L_\gamma[X,W]$ such that $\theta, X, W, A \in N$;

$N_{\nu+1}$ = the smallest $N \prec L_\gamma[X,W]$ such that $N_\nu \cup \{N_\nu\} \subseteq N$;

$N_\lambda = \bigcup_{\nu<\lambda} N_\nu$, if $\lim(\lambda)$.

Let $\pi_\nu : N_\nu \cong \bar{N}_\nu$, where $\bar{N}_\nu$ is transitive. Let $\alpha_\nu = \omega_1 \cap N_\nu$, $\beta_\nu = \pi_\nu''(N_\nu \cap \gamma)$. Then $\alpha_\nu \in \omega_1$, $\alpha_\nu = \pi_\nu(\omega_1)$, and $\langle \alpha_\nu | \nu < \omega_1 \rangle$ is a normal sequence in ω_1. It is clear that $X \cap \alpha_\nu = \pi_\nu(X)$, $A \cap \alpha_\nu = \pi_\nu(A)$, $W\restriction\alpha_\nu = \pi_\nu(W)$, $\bar{N}_\nu = L_{\beta_\nu}[X\cap\alpha_\nu, W\restriction\alpha_\nu]$.

Since $X \in M$, a simple forcing argument shows that there is $\tau < \omega_1$ such that for all $\alpha < \omega_1$, $\alpha \geq \tau \rightarrow X \cap \alpha \in W(\alpha)$. Let τ_0 be the least such. Then τ_0 is $L_\gamma[X,W]$-definable, so $\tau_0 \in N_\nu$ for all $\nu < \omega_1$, which means $\pi_\nu(\tau_0) = \tau_0 < \alpha_\nu$. Hence, for all $\nu < \omega_1$, $X \cap \alpha_\nu \in W(\alpha_\nu)$.

Claim. For all $\nu < \omega_1$, $\beta_\nu \leq \delta(\alpha_\nu)$.

Case 1: $\omega_1 = \omega_1^{L[X\cap\theta, W\restriction\theta]}$. Thus, as $\alpha_\nu > \theta$, α_ν is countable in $L[X\cap\alpha_\nu, W\restriction\alpha_\nu]$.

But α_ν is uncountable in $L_{\beta_\nu}[X\cap\alpha_\nu, W\restriction\alpha_\nu]$ (being "ω_1" there, in fact). Hence $\beta_\nu < \delta(\alpha_\nu)$.

Case 2: $(\forall\eta<\omega_1)(\omega_1^{L[X\cap\eta,W\restriction\eta]}<\omega_1)$. By a simple condensation argument, it follows that $(\forall\eta<\omega_1)(\omega_1$ is inaccessible in $L[X\cap\eta,W\restriction\eta])$. Hence $(\forall\eta<\omega_1)(\omega_2^{L[X\cap\eta,W\restriction\eta]}<\omega_1)$. Suppose α_ν is countable in $L[X\cap\alpha_\nu,W\restriction\alpha_\nu]$. Then, as in Case 1, $\beta_\nu<\delta(\alpha_\nu)$. On the other hand, suppose α_ν is uncountable in $L[X\cap\alpha_\nu, W\restriction\alpha_\nu]$. Since $\alpha_\nu=\pi_\nu(\omega_1)$, it must be the case that $\alpha_\nu = \omega_1^{L[X\cap\alpha_\nu,W\restriction\alpha_\nu]}$. Hence, $(\alpha_\nu^+)^{L[X\cap\alpha_\nu,W\restriction\alpha_\nu]} = \omega_2^{L[X\cap\alpha_\nu,W\restriction\alpha_\nu]}<\omega_1$. Thus, by definition, $\delta(\alpha_\nu)\geq(\alpha_\nu^+)^{L[X\cap\alpha_\nu,W\restriction\alpha_\nu]}$. But $L_{\beta_\nu}[X\cap\alpha_\nu, W\restriction\alpha_\nu]\models(\forall\xi)(|\xi|\leq\alpha_\nu)$. Hence $(\alpha_\nu^+)^{L[X\cap\alpha_\nu,W\restriction\alpha_\nu]}\geq\beta_\nu$, proving $\delta(\alpha_\nu)\geq\beta_\nu$. This proves the claim.

By the claim, $\nu<\omega_1 \rightarrow A\cap\alpha_\nu\in\bar{N}_\nu\subseteq L_{\delta(\alpha_\nu)}[X\cap\alpha_\nu,W\restriction\alpha_\nu]$. But $(\forall\nu<\omega_1)(X\cap\alpha_\nu\in W(\alpha_\nu))$, so we see that $\nu<\omega_1\rightarrow A\cap\alpha_\nu\in\bar{W}_{\alpha_\nu}$. Since $\{\alpha_\nu \mid \nu<\omega_1\}$ is closed and unbounded in ω_1 , we are done. ∎

We shall need to know that $\Diamond^*$ remains true throughout "most" of our iteration.

<u>Lemma 4</u>

Let M be a c.t.m. of ZFC + $(2^\omega=\omega_1)$ + $(2^{\omega_1}=\omega_2)$ + $\Diamond^*$. Suppose $G\subseteq\omega_1^M$ is such that $M[G]$ is a c.t.m. of ZFC + $(2^\omega=\omega_1)$ + $(2^{\omega_1}=\omega_2)$ having the same cardinalities as M . Then $\Diamond^*$ holds in $M[G]$ also.

Proof: The argument is virtually the same as the above. This time, however, instead of W being a generic sequence, we take it to be a $\Diamond^*$ sequence in M . The only difference in the ensuing argument is that we cannot apply a forcing argument to our set X . However, since $X\in M$, there is a closed unbounded set $C\subseteq\omega_1$ in M such that $\alpha\in C\rightarrow X\cap\alpha\in W(\alpha)$, so if we set $\bar{C}=C\cap\{\alpha_\nu \mid \nu<\omega_1\}$, $\alpha\in\bar{C}\rightarrow A\cap\alpha\in\bar{W}_\alpha$, just as before. ∎

Of course, $\Diamond^*$ is simply a much stronger version of the principle $\Diamond$ which we met in the earlier part of this book. Accordingly, lemma 3 can be regarded as a stronger version of lemma III.4. We shall also require the principle $\Diamond$, but in a slightly different form from that we have used previously.

By H_{ω_1} we mean the set of all hereditarily countable sets. For $\alpha < \omega_1$, we set $H_{\omega_1}(\alpha) = H_{\omega_1} \cap V_\alpha$.

$\Diamond'$: There is a sequence $\langle S_\alpha | \alpha < \omega_1 \rangle$ such that each S_α is a countable subset of $H_{\omega_1}(\alpha)$ and, whenever $A \subseteq H_{\omega_1}$ and $\alpha < \omega_1 \rightarrow |A \cap H_{\omega_1}(\alpha)| \leq \omega$, then $\{\alpha \in \omega_1 \mid A \cap H_{\omega_1}(\alpha) = S_\alpha\}$ is stationary in ω_1.

In view of the next result, we shall henceforth understand by $\Diamond$ the principle formulated above as $\Diamond'$.

<u>Lemma 5</u>.

$\Diamond \leftrightarrow \Diamond'$. (Hence $\Diamond^* \rightarrow \Diamond'$.)

Proof: Clearly, $\Diamond' \rightarrow \Diamond$. Now assume $\Diamond$. Then $2^\omega = \omega_1$, so there is $h: \omega_1 \leftrightarrow H_{\omega_1}$. For each $\alpha < \omega_1$, set $S_\alpha = (h''\bar{S}_\alpha) \cap H_{\omega_1}(\alpha)$, where $\langle \bar{S}_\alpha \mid \alpha < \omega_1 \rangle$ realises $\Diamond$. We show that $\langle S_\alpha \mid \alpha < \omega_1 \rangle$ realises $\Diamond'$.

Let $A \subseteq H_{\omega_1}$ be such that $\alpha < \omega_1 \rightarrow |A \cap H_{\omega_1}(\alpha)| \leq \omega$. Let $C \subseteq \omega_1$ be closed and unbounded. We seek an $\alpha \in C$ such that $A \cap H_{\omega_1}(\alpha) = S_\alpha$. Let $\bar{A} = h^{-1}{}'' A$. Then $\{\alpha \in \omega_1 \mid \bar{A} \cap \alpha = S_\alpha\}$ is stationary in ω_1. Let $\bar{C} = C \cap \{\alpha \in \omega_1 \mid h''\alpha \subseteq H_{\omega_1}(\alpha)\ \&\ h''(\bar{A} \cap \alpha) = A \cap H_{\omega_1}(\alpha)\}$, a closed unbounded subset of ω_1. Pick $\alpha \in \bar{C}$ such that $\bar{A} \cap \alpha = \bar{S}_\alpha$. Then $\alpha \in C$ and $A \cap H_{\omega_1}(\alpha) = h''(\bar{A} \cap \alpha) = h''\bar{S}_\alpha = (h''\bar{S}_\alpha) \cap H_{\omega_1}(\alpha) = S_\alpha$. ▌

We shall also require the next result, which is not, it will be observed, a consequence of lemma 4, even if we assume $M \models \Diamond^*$!

Lemma 6

Let M be a c.t.m. of $ZFC + \Diamond$. Let $\mathbb{P}$ be a poset in M such that $M \models$ "$\mathbb{P}$ is ω_1-closed". If G is M-generic on $\mathbb{P}$, then $\Diamond$ holds in $M[G]$.

Proof: Since $\omega_1^M = \omega_1^{M[G]}$, it suffices to show that if $\langle S_\nu \mid \nu < \omega_1 \rangle$ realises $\Diamond$ in M, it also realises $\Diamond$ in $M[G]$. Suppose, therefore, that, in M, there is $p \in \mathbb{P}$ such that $p \Vdash$ "$A \subseteq \check{\omega}_1$ & C is a closed unbounded subset of $\check{\omega}_1$". Working in M, we must find a $q \leq p$ such that for some $\nu < \omega_1$, $q \Vdash$ "$\check{\nu} \in C$ & $A \cap \check{\nu} = \check{S}_\nu$". Now, $\mathbb{P}$ is ω_1-closed, so we can inductively pick a sequence $\langle p_\nu \mid \nu < \omega_1 \rangle$ from $\mathbb{P}$ such that $p_0 \leq p$, $\nu < \tau < \omega_1 \rightarrow p_\tau \leq p_\nu$, and for each $\nu < \omega_1$, there are sets A_ν, $C_\nu \subseteq \nu$ such that $p_\nu \Vdash$ "$A \cap \check{\nu} = \check{A}_\nu$ & $C \cap \check{\nu} = \check{C}_\nu$". Let $\bar{A} = \bigcup_{\nu<\omega_1} A_\nu$ and $\bar{C} = \bigcup_{\nu<\omega_1} C_\nu$. Then $\bar{A}, \bar{C} \subseteq \omega_1$ and $\bar{C}$ is closed and unbounded in ω_1, so there is $\alpha \in \bar{C}$ such that $\bar{A} \cap \alpha = S_\alpha$. Then $p_{\alpha+1} \Vdash$ "$A \cap \check{\alpha} = \check{S}_\alpha$ & $\check{\alpha} \in C$", as required. ▌

Suppose then we are given a c.t.m. M of $ZFC + GCH$. Our eventual aim is to obtain a cardinal absolute generic extension of M, having the same reals as M, in which there are no Souslin trees. As a first step, we make two initial generic extensions of M to get $\Box$ and $\Diamond^*$ to hold. Our previous lemmas enable us to do this without affecting any of our stated requirements. Accordingly, we shall always assume that $\Box$ and $\Diamond^*$ already hold in M. Lemma 4 will tell us that $\Diamond^*$ will remain valid during a large part of our construction, whilst lemma 6 will enable us to use $\Diamond$ even when $\Diamond^*$ is lost.

We now describe the iteration. As usual, it is convenient to formulate everything in boolean terms. Since we are forcing with Souslin trees, we are thus led naturally to the concept of a <u>Souslin algebra</u>.

A Souslin algebra is a BA $\mathbb{B}$ such that there is a dense set $T \subseteq \mathbb{B} - \{\mathbb{0}\}$ which is, under $\leq_{\mathbb{B}}$, a Souslin tree.

If $\mathbb{B}$ is a Souslin algebra and T is a dense subset of $\mathbb{B} - \{\mathbb{0}\}$ which is a Souslin tree under $\leq_{\mathbb{B}}$, we say that T Souslinises $\mathbb{B}$.

The next lemma characterises Souslin algebras in purely boolean terms, at least under the assupption $2^{\omega} = \omega_1$, which we shall have throughout.

Lemma 7

Assume $2^{\omega} = \omega_1$. Let $\mathbb{B}$ be a BA. $\mathbb{B}$ is a Souslin algebra iff (i) - (iv) hold, where:

(i) $|\mathbb{B}| = \omega_1$;

(ii) $\mathbb{B}$ is (ω_1, ∞)-distributive;

(iii) $\mathbb{B}$ satisfies c.c.c.

(iv) $\mathbb{B}$ is atomless.

Proof: Suppose $\mathbb{B}$ is a Souslin algebra. Let T Souslinise $\mathbb{B}$. Since T is dense in $\mathbb{B}$, $\mathbb{B}$ satisfies c.c.c. Again, the density of T means that T generates $\mathbb{B}$ as a BA, which, in view of the fact that $\mathbb{B}$ satisfies c.c.c. means that $|\mathbb{B}| \leq |T|^{\omega} = \omega_1^{\omega} = \omega_1$. Finally, by lemma VII.3, we see that $\mathbb{B}$ is ω_1-dense, and hence, by lemma VII.1, (ω_1, ∞)-distributive.

Conversely, suppose $\mathbb{B}$ satisfies (i) - (iv). Construct a subset T of $\mathbb{B}$ such that $\langle T, \leq_{\mathbb{B}} \rangle$ is a tree, by induction on the levels as follows. Let $\langle b_{\alpha} | \alpha < \omega_1 \rangle$ enumerate $\mathbb{B}$. Set $T_0 = \{\mathbb{1}_{\mathbb{B}}\}$. If T_{α} is defined, obtain $T_{\alpha+1}$ as follows. Let $b \in T_{\alpha}$. Let $b \wedge b_{\gamma}$ and $b \wedge - b_{\gamma}$ extend b on $T_{\alpha+1}$, where γ is least such that neither of these meets is $\mathbb{0}$ and yet b_{γ} has not yet been so considered. If $\lim(\alpha)$ and T_{β}, $\beta < \alpha$, are defined, let T_{α} consist of all non-zero meets of the form $\bigwedge_{\beta<\alpha} p_\beta$, where $\beta<\gamma<\alpha \to p_\beta \in T_\beta$ & $p_\gamma \leq p_\beta$. By conditions (ii) - (iv), it is easily seen that T is in fact a Souslin tree, dense in $\mathbb{B}$. (We need clause (ii) in order to see that every point lies on an α-branch which is extended for each limit α . And by clause (iii), T will, in particular, not have height greater than ω_1.) ▮

Remark. Without the assumption of CH , lemma 7 holds for any BA of cardinality ω_1 , of course. Hence the existence of a BA of cardinality ω_1 which satisfies conditions (ii) - (iv) of lemma 7 is neither provable nor refutable in ZFC .

The following lemma is fairly trivial, but well worth a special mention.

Lemma 8

Let $\mathbb{B}$ be a Souslin algebra, and let $\underset{\sim}{T}$ Souslinise $\mathbb{B}$. Then:

(i) $\mathbb{1}$ is the root of $\underset{\sim}{T}$;

(ii) If x,y are $\underset{\sim}{T}$-incomparable elements of T , then $x \wedge y = \mathbb{0}$;

(iii) If $\{x_n | n<\omega\} \subseteq T$ and $\bigwedge_{n<\omega} x_n > \mathbb{0}$, then $\bigwedge_{n<\omega} x_n \in T$.

Proof: (i), (ii) are trivial.

(iii) uses the uniqueness of limit points in $\underset{\sim}{T}$. ▮

The next lemma shows that if $\mathbb{B}$ is a Souslin algebra, then any two Souslinisations of $\mathbb{B}$ are essentially the same.

Lemma 9

Let $\mathbb{B}$ be a Souslin algebra, and let T,T' Souslinise $\mathbb{B}$. Then there is a closed unbounded set $A \subseteq \omega_1$ such that $\alpha \in A \to T_\alpha = T'_\alpha$.

Proof: As T' is dense in $\mathbb{B}$, for each $x \in T$ we may pick a pairwise disjoint set $Y \subseteq T'$ such that $x = \bigvee Y$. Since T' is Souslin, Y is countable, so we can find a $\beta < \omega_1$ such that $x = \bigvee\{y \in T'|\beta \mid y \leq x\}$. Similarly, for each $y \in T'$ we can find a $\beta < \omega_1$ such that $y = \bigvee\{x \in T|\beta \mid x \leq y\}$.

Using our above observations, we may define functions η, η' on ω_1 to ω_1 as follows:

$\eta(\alpha)$ = the least $\beta > \alpha$ such that $x \in T_\alpha \rightarrow x = \bigvee\{y \in T'|\beta \mid y \leq x\}$

$\eta'(\alpha)$ = the least $\beta > \alpha$ such that $y \in T'_\alpha \rightarrow y = \bigvee\{x \in T|\beta \mid x \leq y\}$.

Define a normal function $\beta : \omega_1 \rightarrow \omega_1$ by induction, thus:

$\beta(0) = 0$, $\beta(2\nu+1) = \eta \circ \beta(2\nu)$, $\beta(2\nu+2) = \eta' \circ \beta(2\nu+1)$,

$\beta(\delta) = \sup_{\nu<\delta} \beta(\nu)$ if $\lim(\delta)$.

Let $A = \{\beta(\lambda) \mid \lambda < \omega_1 \ \& \ \lim(\lambda)\}$. We show that A is as required. Let $\lambda < \omega_1$ be an arbitrary limit ordinal. For each $x \in T_{\beta(\lambda)}$, let $Y(x) = \{y \in T'_{\beta(\lambda)} \mid y \leq x\}$, and for each $y \in T'_{\beta(\lambda)}$, let $X(y) = \{x \in T_{\beta(\lambda)} \mid x \leq y\}$. Then we claim that:

(*) $x \in T_{\beta(\lambda)} \rightarrow x = \bigvee Y(x)$ and (**) $y \in T'_{\beta(\lambda)} \rightarrow y = \bigvee X(y)$.

We verify (*), the proof of (**) being entirely similar. Let $x \in T_{\beta(\lambda)}$. For each $\nu < \beta(\lambda)$, let x_ν be that $x' \in T_\nu$ such that $x \leq x'$, and set $Y_\nu = \{y \in T'_{\eta(\nu)} \mid y \leq x_\nu\}$. By definition of η, $x_\nu = \bigvee Y_\nu$ for each $\nu < \beta(\lambda)$. If $\langle y_{\nu,n} \mid n < \omega\rangle$ enumerates Y_ν for each $\nu < \beta(\lambda)$, then we see that $x = \bigwedge_{\nu<\beta(\lambda)} x_\nu = \bigwedge_{\nu<\beta(\lambda)} \bigvee_{n<\omega} y_{\nu,n} = \bigvee_{f\in\omega^{\beta(\lambda)}} \bigwedge_{\nu<\beta(\lambda)} y_{\nu,f(\nu)}$ —— ⊛

Now, for any $f \in \omega^{\beta(\lambda)}$, $\bigwedge_{\nu<\beta(\lambda)} y_{\nu,f(\nu)} > \mathbb{0} \rightarrow \bigwedge_{\nu<\beta(\lambda)} y_{\nu,f(\nu)} \in T'_{\beta(\lambda)}$. And by ⊛, $\bigwedge_{\nu<\beta(\lambda)} y_{\nu,f(\nu)} \leq x$ in this case.
Hence, by ⊛ and the definition of $Y(x)$, $x = \bigvee Y(x)$, proving (*).

Finally, let $x \in T_{\beta(\lambda)}$. Pick $y \in Y(x)$, $x' \in X(y)$. (By

(*), $Y(x) \neq \emptyset$, and by (**), $X(y) \neq \emptyset$, so this causes no problem.) Then $x, x' \in T_{\beta(\lambda)}$ and $x' \leq y \leq x$, so we must have $x' = y = x$. Hence $x = y \in T'_{\beta(\lambda)}$, proving that $T_{\beta(\lambda)} \subseteq T'_{\beta(\lambda)}$. Similarly, $T'_{\beta(\lambda)} \subseteq T_{\beta(\lambda)}$, so we are done. ∎

We are now almost ready to formulate and prove our main iteration lemma for forcing with Souslin algebras. At limit stages we shall use a modified inverse limit construction. The idea is this. We force with Souslin algebras at each stage, and carry along an arbitrary Souslinisation for each stage. (Lemma 9 will make irrelevant the actual choice of this Souslinisation.) At a limit stage we first of all form the inverse limit, using our Souslinisations, and then "thin down" this limit to a Souslin tree which is still large enough to completely embed all of the previous Souslin algebras in its boolean algebra. In view of our description of the inverse limit construction given in the previous chapter, we should expect to have some property corresponding to "ω_1-happiness", which holds between any two algebras in the iteration sequence. The notion which we use is that of a "nice" subalgebra, or "nice" embedding.

Let $\mathbb{B}, \mathbb{B}'$ be Souslin algebras. $\mathbb{B}$ is a <u>nice</u> subalgebra of $\mathbb{B}'$ iff $\mathbb{B}$ is a complete subalgebra of $\mathbb{B}'$ and there are Souslinisations T, T' of $\mathbb{B}, \mathbb{B}'$ (respectively) such that $\{\alpha \in \omega_1 \mid T_\alpha = h'' T'_\alpha\}$ contains a closed unbounded set, where $h : \mathbb{B}' \to \mathbb{B}$ is the basic projection. The notion of a <u>nice</u> embedding is defined analogously in the obvious way.

The next lemma removes from the above definition the ostensible dependence upon the choice of Souslinisations.

<u>Lemma 10</u>

Let $\mathbb{B}, \mathbb{B}'$ be Souslin algebras, and suppose $\mathbb{B}$ is a nice subalgebra of $\mathbb{B}'$. If T, T' are arbitrary Souslinisations of $\mathbb{B}$, $\mathbb{B}'$ (respectively), then $\{\alpha \in \omega_1 \mid T_\alpha = h'' T'_\alpha\}$ contains a closed unbounded set, where $h : \mathbb{B}' \to \mathbb{B}$ is the basic projection.

Proof: By lemma 9. ▌

The next lemma shows that the notion of a nice subalgebra is transitive, which we shall clearly need if this is to play a similar role to happiness.

<u>Lemma 11</u>

Let $\mathbb{B}_0, \mathbb{B}_1, \mathbb{B}_2$ be Souslin algebras. Suppose $\mathbb{B}_0$ is a nice subalgebra of $\mathbb{B}_1$ and $\mathbb{B}_1$ is a nice subalgebra of $\mathbb{B}_2$. Then $\mathbb{B}_0$ is a nice subalgebra of $\mathbb{B}_2$.

Proof: Let $h_{ij} : \mathbb{B}_i \to \mathbb{B}_j$, $0 \leq j < i \leq 2$, be the basic projections. Let T^i be any Souslinisation of $\mathbb{B}_i$, $0 \leq i \leq 2$. Let A_{01}, A_{12} be closed unbounded subsets of ω_1 such that $\alpha \in A_{01} \to T^0_\alpha = h_{10}'' T^1_\alpha$ and $\alpha \in A_{12} \to T^1_\alpha = h_{21}'' T^2_\alpha$. Let $A_{02} = A_{01} \cap A_{12}$. But clearly, $h_{10} \circ h_{21} = h_{20}$, so we are done. ▌

<u>Lemma 12</u> (Iteration Lemma)

Assume $(2^\omega = \omega_1) + (2^{\omega_1} = \omega_2) + \Diamond + \Box$. Let σ be a function such that for all $\mathbb{B}$, x, if $\mathbb{B}$ is a Souslin algebra, or if $\mathbb{B} = \mathbb{1}$, then $\sigma(\mathbb{B},x)$ is a Souslin algebra of which, in the case $\mathbb{B} \neq \mathbb{1}$, $\mathbb{B}$ is a nice subalgebra. Then there is a sequence $\langle \mathbb{B}_\nu \mid \nu < \omega_2 \rangle$ such that:

(i) $\mathbb{B}_0 = \mathbb{1}$;

(ii) $\mathbb{B}_\nu$ is a Souslin algebra, for all $\nu > 0$;

(iii) $\mathbb{B}_{\nu+1} = \sigma(\mathbb{B}_\nu, \langle \mathbb{B}_\tau \mid \tau < \nu \rangle)$ for all ν;

(iv) if $0 < \tau < \nu$, $\mathbb{B}_\tau$ is a nice subalgebra of $\mathbb{B}_\nu$.

Proof: We construct the sequence $\langle \mathbb{B}_\nu \mid \nu < \omega_2 \rangle$ by induction. As we proceed, we construct also the following sequences:

(a) $h_{\nu\tau} : \mathbb{B}_\nu \to \mathbb{B}_\tau$ $(\nu \geq \tau)$, the basic projections;

(b) $T^\nu \subseteq \mathbb{B}_\nu$, an arbitrary Souslinisation of $\mathbb{B}_\nu$;

(c) $\ell_\nu : \mathbb{B}_\nu \cong \tilde{\mathbb{B}}_\nu$, where $\tilde{\mathbb{B}}_\nu$ has domain ω_1 but $\ell_\nu, \tilde{\mathbb{B}}_\nu$ are otherwise arbitrary;

(d) $\tilde{T}^\nu = \ell_\nu "T^\nu$; $\tilde{h}_{\nu\tau} = \ell_\tau \circ h_{\nu\tau} \circ \ell_\nu^{-1}$ $(\nu \geq \tau)$;

(e) $C_{\nu\tau}$ $(\nu \leq \tau)$, an arbitrary closed unbounded set in ω_1 such that for all $\alpha \in C_{\nu\tau}$, $\tilde{T}^\nu | \alpha = \tilde{T}^\nu \cap \alpha$, $\tilde{T}^\tau | \alpha = \tilde{T}^\tau \cap \alpha$, $T^\nu_\alpha = h_{\tau\nu} "T^\tau_\alpha$

We let $\leq_\nu$, $\leq^\sim_\nu$ denote the boolean ordering of $\mathbb{B}_\nu$, $\tilde{\mathbb{B}}_\nu$, respectively.

Let $\langle A_\lambda \mid \lambda < \omega_2 \;\&\; \lim(\lambda) \rangle$ realise $\square$ and let $\langle S_\alpha \mid \alpha < \omega_1 \rangle$ realise $\diamondsuit$ (in the form stated earlier).

<u>Case 1</u>. Stage 0.

Set $\mathbb{B}_0 = \mathbf{2}$.

<u>Case 2</u>. Stage $\nu + 1$.

Set $\mathbb{B}_{\nu+1} = \sigma(\mathbb{B}_\nu, \langle \mathbb{B}_\tau \mid \tau < \nu \rangle)$. Condition (iv) holds by lemma 11 and induction.

<u>Case 3</u>. Stage λ, $\lambda < \omega_2$, $\lim(\lambda)$, $\mathrm{cf}(\lambda) = \omega$.

Let $\theta = \mathrm{otp}(A_\lambda)$, and let $\langle \lambda(\nu) \mid \nu < \theta \rangle$ be the normal enumeration of A_λ. Notice that $\theta \leq \lambda$ and $\theta < \omega_1$. We define $\mathbb{B}_\lambda$ from $\langle \mathbb{B}_{\lambda(\nu)} \mid \nu < \theta \rangle$, or, more precisely, from $\langle \tilde{T}^{\lambda(\nu)} \mid \nu < \theta \rangle$.

Set $C = [\bigcap_{\nu<\tau<\theta}(C_{\lambda(\nu),\lambda(\tau)}-\theta)] \cup \{0\}$. Let $\langle c_\nu \mid \nu<\omega_1\rangle$ be the normal enumeration of C . Notice that $c_0 = 0$ and $c_1 \geq \theta$.

For all $\alpha < \omega_1$, $\tilde{T}^{\lambda(\nu)}_{c_\alpha} = \tilde{h}_{\lambda(\tau),\lambda(\nu)}{}''\tilde{T}^{\lambda(\tau)}_{c_\alpha}$, for $\nu < \tau < \theta$, so we may define

$$T^* = \{x = \langle x_\nu \mid \nu < \theta\rangle \mid (\exists\gamma < \omega_1)(\nu < \theta \to x_\nu \in \tilde{T}^{\lambda(\nu)}_{c_\gamma} \ .\&.$$
$$\nu < \tau < \theta \to x_\nu = \tilde{h}_{\lambda(\tau),\lambda(\nu)}(x_\tau))\}$$

Partially-order T^* by $x \leq^* y \leftrightarrow (\forall\nu < \theta)(x_\nu \tilde{\leq}_{\lambda(\nu)} y_\nu)$.

It is easily seen that $\langle T^*, \leq^*\rangle$ is a tree and that $x \in T^*_\alpha \leftrightarrow (\forall\nu < \theta)(x_\nu \in \tilde{T}^{\lambda(\nu)}_{c_\alpha})$.

<u>Crucial Fact</u> (C.F.) There is a subtree $\hat{T}$ of T^* such that:

(1) $\hat{T}$ is a Souslin tree;

(2) if $x \in \hat{T}$, $y \in \tilde{T}^{\lambda(\nu)}$, $y \tilde{\leq}_{\lambda(\nu)} x_\nu$, there is $x' \in \hat{T}$ such that $x' \leq^* x$, $x'_\nu = y$.

We leave the verification of the C.F. until later, and complete the definition of $\mathbb{B}_\lambda$. (But note the similarity between condition (2) and the notion of a "neat cover".)

Let $\mathbb{B} = BA(\hat{T})$. Define maps $\eta_\nu : \tilde{\mathbb{B}}_{\lambda(\nu)} \to \mathbb{B}$, $\nu < \theta$, by $\eta_\nu(b) = \{x \in \hat{T} \mid x_\nu \tilde{\leq}_{\lambda(\nu)} b\}$.

Using C.F. (2), it is easily checked that η_ν is a complete embedding of $\tilde{\mathbb{B}}_{\lambda(\nu)}$ into $\mathbb{B}$. (In particular, the mapping η_ν is one-one.)

Pick $\mathbb{B}_\lambda$, $\bar{\eta}$ to make the following diagrams commute for $\nu < \tau < \theta$

$$\begin{array}{ccccc}
\eta_\nu{}''\tilde{\mathbb{B}}_{\lambda(\nu)} & \subseteq & \eta_\tau{}''\tilde{\mathbb{B}}_{\lambda(\tau)} & \subseteq & \mathbb{B} \\
\cong\Big\uparrow \eta_\nu & & \cong\Big\uparrow \eta_\tau & & \cong\Big\uparrow \bar{\eta} \\
\mathbb{B}_{\lambda(\nu)} & & \tilde{\mathbb{B}}_{\lambda(\tau)} & & \\
\cong\Big\uparrow \ell_{\lambda(\nu)} & & \cong\Big\uparrow \ell_{\lambda(\tau)} & & \\
\mathbb{B}_{\lambda(\nu)} & \subseteq & \mathbb{B}_{\lambda(\tau)} & \subseteq & \mathbb{B}_\lambda .
\end{array}$$

Define $\eta : \hat{T} \to \mathbb{B}_\lambda$ by $\eta(x) = \bigwedge_{\nu<\theta} \ell^{-1}_{\lambda(\nu)}(x_\nu)$. Then $\eta : \langle \hat{T}, \leq^* \rangle \cong \langle \eta''\hat{T}, \leq_\lambda \rangle$, and $\eta''\hat{T}$ Souslinises $\mathbb{B}_\lambda$. We may as well set $T^\lambda = \eta''\hat{T}$ for definiteness.

Let $h_\nu : \mathbb{B}_\lambda \to \mathbb{B}_{\lambda(\nu)}$, $\nu < \theta$, be the basic projections. Clearly, $h_\nu(\eta(x)) = \ell^{-1}_{\lambda(\nu)}(x_\nu)$ for each $x \in \hat{T}$. Hence for each $\nu < \theta$, $\alpha < \omega_1$ & $\alpha = c_\alpha \to T^{\lambda(\nu)}_\alpha = h_\nu{}''T^\lambda_\alpha$. Thus $\mathbb{B}_{\lambda(\nu)}$ is a nice subalgebra of $\mathbb{B}_\lambda$, each $\nu < \theta$. By lemma 11, therefore, $\mathbb{B}_\nu$ is a nice subalgebra of $\mathbb{B}_\lambda$ for all $\nu < \lambda$. This completes the construction of $\mathbb{B}_\lambda$, and it remains to verify the C.F.

Claim 1. Suppose $\langle \nu, y \rangle$, $\langle \tau, x \rangle$ are such that $\nu \leq \tau < \theta$ and for some $c \in C$, $y \in \tilde{T}^{\lambda(\nu)}_c$ and $x \in \tilde{T}^{\lambda(\tau)}$ and $y \leq_{\widetilde{\lambda(\nu)}} \tilde{h}_{\lambda(\tau),\lambda(\nu)}(x)$. Then there is $x' \in \tilde{T}^{\lambda(\tau)}_c$ such that $x' \leq_{\widetilde{\lambda(\tau)}} x$ and $\tilde{h}_{\lambda(\tau),\lambda(\nu)}(x') = y$.

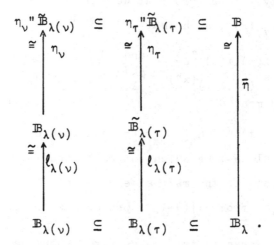

By assumption $\ell^{-1}_{\lambda(\nu)}(y) \leq_{\lambda(\nu)} h_{\lambda(\tau),\lambda(\nu)}(\ell^{-1}_{\lambda(\tau)}(x))$. Hence $\ell^{-1}_{\lambda(\nu)}(y) \wedge \ell^{-1}_{\lambda(\tau)}(x) >_{\lambda(\tau)} \mathbb{0}$. Since $c \in C$, it is easily seen that x lies above $\tilde{T}^{\lambda(\tau)}_c$ (in terms of $<_{\widetilde{\lambda(\tau)}}$). Hence, by density, it follows that we can find $x" \in T_c^{\lambda(\tau)}$, $x" \leq_{\lambda(\tau)} \ell^{-1}_{\lambda(\tau)}(x)$, such that $x" \wedge \ell^{-1}_{\lambda(\nu)}(y) >_{\lambda(\tau)} \mathbb{0}$. Since $T_c^{\lambda(\nu)} = h_{\lambda(\tau),\lambda(\nu)}{}"T_c^{\lambda(\tau)}$ and $T_c^{\lambda(\nu)}$ is pairwise disjoint in $\mathbb{B}_{\lambda(\nu)}$, $h_{\lambda(\tau),\lambda(\nu)}(x") = \ell^{-1}_{\lambda(\nu)}(y)$. Thus $x' = \ell_{\lambda(\tau)}(x")$ is as claimed.

We construct $\hat{T} = \cup_{\alpha<\omega_1} \hat{T}_\alpha$ by induction on α so that $\hat{T}|\alpha$ is a normal α-tree which satisfies condition (2) of the C.F. $\hat{T}_o$ consists of the maximal element of T^*, say $\bar{x}$. To obtain $\hat{T}_{\alpha+1}$ from $\hat{T}|(\alpha+1)$, proceed as follows.

Claim 2. For each triple $\langle \nu,y,x\rangle$ such that $\nu < \theta$, $x \in \hat{T}_\alpha$, $y \in \tilde{T}^{\lambda(\nu)}_{c_{\alpha+1}}$, and $y \tilde{\leq}_{\lambda(\nu)} x_\nu$, there is $s \in T^*_{\alpha+1}$, $s \leq^* x$, such that $s_\nu = y$.

Let $\langle \nu(n) \mid n < \omega\rangle$ be cofinal in θ, with $\nu(0) = \nu$. By claim 1, we can inductively pick, for each $n < \omega$, $y_n \in \tilde{T}^{\lambda(\nu(n))}_{c_{\alpha+1}}$, such that $y_o = y$, $y_n \tilde{\leq}_{\lambda(\nu(n))} x_{\nu(n)}$, and $y_n = \tilde{h}_{\lambda(\nu(n+1)),\lambda(\nu(n))}(y_{n+1})$. Define $s : \theta \to V$ by $s_\tau = \tilde{h}_{\lambda(\nu(n)),\lambda(\tau)}(y_n)$ where n is large enough for $\nu(n) \geq \tau$. By construction, this well-defines s, and s is clearly as required.

For each triple $\langle \nu,y,x\rangle$ as above, let $s(\nu,y,x)$ be <u>one</u> such s as claim 2 guarantees. Let $\hat{T}_{\alpha+1}$ consist of all the $s(\nu,y,x)$ for all possible $\langle \nu,y,x\rangle$. The actual choice of $s(\nu,y,x)$ is irrelevant except when $\alpha = 0$ and:

(*) For all $\nu < \theta$, $y \in \tilde{T}^{\lambda(\nu)}_{c_1}$, there is a pair $\langle \tau,z\rangle \in S_{c_1}$ such that $\tau \geq \nu$, $\tau < \theta$, $z \in \tilde{T}^{\lambda(\tau)}|c_1$, and $y \tilde{\leq}_{\lambda(\nu)} \tilde{h}_{\lambda(\tau),\lambda(\nu)}(z)$.

In this case, we select $s(\nu,y,\bar{x})$ in such a way that for some $\langle \tau, z \rangle \in S_{c_1}$, $z \in \tilde{T}^{\lambda(\tau)}|c_1$ and $s_\tau \leq^{\sim}_{\lambda(\tau)} z$. This is possible by (*) itself and claims 1 and 2.

Finally, suppose $\lim(\alpha)$ and that $\hat{T}|\alpha$ is defined. $\hat{T}_\alpha$ is obtained as follows.

Claim 3. Let U be the set of all triples $\langle \nu, y, x \rangle$ such that $x \in \hat{T}|\alpha$, $y \in \tilde{T}^{\lambda(\nu)}_{c_\alpha}$, and $y \leq^{\sim}_{\lambda(\nu)} x_\nu$. For each $\langle \nu, y, x \rangle \in U$ there is $s \in T^*_\alpha$ such that $s \leq^* x$, $s_\nu = y$, and $\{x' \in \hat{T}|\alpha \mid s \leq^* x'\}$ is an α-branch of $\hat{T}|\alpha$.

Let $\langle \nu(n) \mid n < \omega \rangle$ be cofinal in θ with $\nu(0) = \nu$. Let $\langle \tau(n) \mid n < \omega \rangle$ be cofinal in α with $x \in \hat{T}_{\tau(0)}$. By claim 1 and the fact that $\hat{T}|\alpha$ satisfies C.F. (2) we can inductively pick, for $n < \omega$, y_n, x_n such that $y_0 = y$, $x_0 = x$, $\langle \nu(n), y_n, x_n \rangle \in U$, $\tilde{h}_{\lambda(\nu(n+1)),\lambda(\nu(n))}(y_{n+1}) = y_n$, $x_{n+1} \leq^* x_n$, $x_n \in \hat{T}_{\tau(n)}$. Define s by $s_\tau = \tilde{h}_{\lambda(\nu(n)),\lambda(\tau)}(y_n)$ where $\nu(n) \geq \tau$, all $\tau < \theta$. Then s , capping the branch $\langle x_n \mid n < \omega \rangle$ of $\hat{T}|\alpha$, is as required.

To obtain $\hat{T}_\alpha$, for each $\langle \nu, y, x \rangle \in U$ we pick one such $s(\nu, y, x)$ in T^*_α as claim 3 guarantees, and put this into $\hat{T}_\alpha$. The actual choice of $s(\nu, y, x)$ is irrelevant except in the following case:

(**) Suppose $c_\alpha = \alpha$ and $S_\alpha \subseteq \hat{T}|\alpha$, and for every $\langle \nu, y, x \rangle \in U$ there is an $x' \in S_\alpha$ such that $\langle \nu, y, x' \rangle \in U$ and $x' \leq^* x$.

In this event, we select $s(\nu, y, x)$ so that $s(\nu, y, x) \leq^* x'$ for some $x' \in S_\alpha$. This is possible by the condition (**) itself, of course.

Let $\hat{T} = \bigcup_{\alpha < \omega_1} \hat{T}_\alpha$. Clearly, $\hat{T}$ is a normal ω_1-tree satisfying C.F. (2). We show that $\hat{T}$ is Souslin.

Let X be a maximal antichain of $\hat{T}$. Let $D = \{y \in \hat{T} \mid (\exists x \in X)(y \leq^* x)\}$, a dense initial section of $\hat{T}$. For each $\alpha < \omega_1$, set $D_\alpha = D \cap (\hat{T}|\alpha)$, $X_\alpha = X \cap (\hat{T}|\alpha)$. Notice that for all α, $D_\alpha = \{y \in \hat{T}|\alpha \mid (\exists x \in X_\alpha)(y \leq^* x)\}$.

Let $\nu < \theta$. For each $y \in \tilde{T}^{\lambda(\nu)}$ and each $x \in \hat{T}$ with $y \leq_{\tilde{\lambda}(\nu)} x_\nu$, let $Y(y,x)$ be a maximal pairwise incomparable (in $\tilde{T}^{\lambda(\nu)}$) set of $y' \leq_{\tilde{\lambda}(\nu)} y$ such that for some $x' \in D$, $x' \leq^* x$, we have $x'_\nu = y'$. Since D is dense and $\hat{T}$ satisfies C.F. (2), $y = \bigvee Y(y,x)$.

Let K_ν be the set of all $\alpha < \omega_1$ such that $c_\alpha = \alpha > 0$ and whenever $y \in \tilde{T}^{\lambda(\nu)}|\alpha$, $x \in \hat{T}|\alpha$, and $y \leq_{\tilde{\lambda}(\nu)} x_\nu$, then $Y(y,x) \subseteq \hat{T}|\beta$ for some $\beta < \alpha$. Clearly, K_ν is closed and unbounded in ω_1.

<u>Claim 4</u>. If $\alpha \in K_\nu$ and $y \in \tilde{T}_\alpha^{\lambda(\nu)}$, then D_α is dense for $\{x \in \hat{T}|\alpha \mid y \leq_{\tilde{\lambda}(\nu)} x_\nu\}$. [By "$D$ is dense <u>for</u> Z", we mean that $D \cap Z$ is a dense subset of Z.]

Let $x \in \hat{T}|\alpha$, $y \leq_{\tilde{\lambda}(\nu)} x_\nu$. Pick $y' \in \tilde{T}^{\lambda(\nu)}|\alpha$ such that $y \leq_{\tilde{\lambda}(\nu)} y' \leq_{\tilde{\lambda}(\nu)} x_\nu$. Since $y' = \bigvee Y(y',x)$, there is $y'' \in Y(y',x)$ such that $y \leq_{\tilde{\lambda}(\nu)} y''$. By definition of $Y(y',x)$, there is $x' \in D_\alpha$ such that $x' \leq^* x$ and $y'' = x'_\nu$. That is what we required.

Let $K = \bigcap_{\nu<\theta} K_\nu$, a closed unbounded subset of ω_1. Let $\alpha \in K$ be such that, by $\diamondsuit$, $S_\alpha = D \cap V_\alpha$. Since $c_\alpha = \alpha$, $\tilde{T}^{\lambda(\nu)}|\alpha = \tilde{T}^{\lambda(\nu)} \cap \alpha$ for all $\nu < \theta$, so, recalling that $\theta \leq c_1$, we see that $\hat{T}|\alpha = \hat{T} \cap V_\alpha$. Thus $S_\alpha = D \cap (\hat{T}|\alpha) = D_\alpha$. By claim 4, therefore, the special case (**) applied in defining $\hat{T}_\alpha$. Hence, each $x \in \hat{T}_\alpha$ extends an element of D_α. This means that each $x \in \hat{T}_\alpha$ extends an element of X_α, whence X_α is a maximal antichain of $\hat{T}$. Hence $X = X_\alpha$, which is countable.

Case 4. Stage λ, $\lambda < \omega_2$, $\lim(\lambda)$, $cf(\lambda) = \omega_1$.

Set $\mathbb{B}_\lambda = \bigcup_{\nu<\lambda} \mathbb{B}_\nu$. We must prove that $\mathbb{B}_\lambda$ is a Souslin algebra and that for every $\nu < \lambda$, $\mathbb{B}_\nu$ is a nice subalgebra of $\mathbb{B}_\lambda$. Notice that we do not know automatically that $\mathbb{B}_\lambda$ is even complete. However, providing we can find a Souslinisation of $\mathbb{B}_\lambda$, this will not matter, because then $\mathbb{B}_\lambda$ will satisfy c.c.c., which itself implies the completeness of $\mathbb{B}_\lambda$.

Let $\langle \lambda(\nu) \mid \nu < \omega_1 \rangle$ enumerate A_λ. Define closed unbounded sets $B_\nu \subseteq \omega_1$, $\nu < \omega_1$, as follows. $B_0 = \omega_1$; $B_{\nu+1} = B_\nu \cap [\bigcap_{\tau<\nu} C_{\lambda(\tau),\lambda(\nu)}] \cap [\bigcap_{\tau\le\nu} C_{\lambda(\tau),\lambda(\nu+1)}]$; $B_\gamma = \bigcap_{\nu<\gamma} B_\nu$, if $\lim(\gamma)$. Since $B_0 \supseteq B_1 \supseteq \dots$, we can pick a normal sequence $\langle \beta_\nu \mid \nu < \omega_1 \rangle$ such that $\beta_0 = 0$ and $(\forall \nu < \omega_1)(\beta_\nu \in B_\nu)$. Then, for all $\gamma < \omega_1$:

$$\text{(i)} \quad \nu < \gamma \rightarrow \tilde{T}^{\lambda(\nu)} | \beta_\gamma = \tilde{T}^{\lambda(\nu)} \cap \beta_\gamma ;$$

$$\text{(ii)} \quad \nu \le \tau < \gamma \rightarrow T^{\lambda(\nu)}_{\beta_\gamma} = h_{\lambda(\tau),\lambda(\nu)}{}'' T^{\lambda(\tau)}_{\beta_\gamma} ;$$

$$\text{(iii)} \quad \nu \le \gamma \rightarrow T^{\lambda(\nu)}_{\beta_{\gamma+1}} = h_{\lambda(\gamma+1),\lambda(\nu)}{}'' T^{\lambda(\gamma+1)}_{\beta_{\gamma+1}} .$$

Define sets $T_\nu \subseteq \mathbb{B}_\lambda$ by induction on $\nu < \omega_1$ as follows, $T_0 = \{\mathbb{1}\}$; $T_{\nu+1} = T^{\lambda(\nu+1)}_{\beta_{\nu+1}}$; $T_\alpha = \{\bigwedge^{\mathbb{B}_{\lambda(\alpha)}}_{\nu<\alpha} x_\nu > \mathbb{0} \mid \nu < \alpha \rightarrow x_\nu \in T^{\lambda(\nu)}_{\beta_\alpha}$.&. $\nu \le \tau < \alpha \rightarrow x_\nu = h_{\lambda(\tau),\lambda(\nu)}(x_\tau)\}$, if $\lim(\alpha)$. Set $T = \bigcup_{\nu<\omega_1} T_\nu$. By (ii), (iii) above, $\langle T, \le_\lambda \rangle$ is a tree with α'th level T_α as just defined. In fact, T is easily seen to be a normal ω_1-tree, dense in $\mathbb{B}_\lambda$ (as $\mathbb{B}_\lambda = \bigcup_{\nu<\omega_1} \mathbb{B}_{\lambda(\nu)}$.)

We show that T is Souslin.

Let X be a maximal antichain of T, and for every $\alpha < \omega_1$, set $X_\alpha = X \cap (T|\alpha)$. Let $K_0 = \{\alpha \in \omega_1 \mid \lim(\alpha)\ \&\ X_\alpha$ is a maximal antichain of $T|\alpha\}$, a closed unbounded subset of ω_1. Let $D = \{y \in T \mid (\exists x \in X)(y \leq_\lambda x)\}$, and for every $\alpha < \omega_1$, set $D_\alpha = \{y \in T|\alpha \mid (\exists x \in X_\alpha)(y \leq_\lambda x)\}$. Then D is a dense initial section of T and for any $\alpha \in K_0$, D_α is a dense initial section of $T|\alpha$.

Let $K_1 = \{\alpha \in \omega_1 \mid \beta_\alpha = \alpha$.&. $\lambda(\alpha)$ is a limit point of A_λ .&. $\alpha = \mathrm{otp}(A_{\lambda(\alpha)})\}$. It is easily seen that K_1 is a closed unbounded subset of ω_1.

<u>Claim 5</u>. For $\nu < \omega_1$, $E_\nu = \{y \in T^{\lambda(\nu)} \mid (\exists \tau < \omega_1)(\exists z \in T^{\lambda(\tau)})[\tau \geq \nu$.&. $z \in D$.&. $h_{\lambda(\tau),\lambda(\nu)}(z) = y]\}$ is dense in $\mathbb{B}_{\lambda(\nu)}$.

Let $y \in T^{\lambda(\nu)}$. Pick $\gamma < \omega_1$, $\gamma > \nu$, so that $y \in T^{\lambda(\nu)}|\beta_{\gamma+1}$, and let $y' \leq_{\lambda(\nu)} y$, $y' \in T^{\lambda(\nu)}_{\beta_{\gamma+1}}$. For some $z' \in T^{\lambda(\gamma+1)}_{\beta_{\gamma+1}}$, $y' = h_{\lambda(\gamma+1),\lambda(\nu)}(z')$. Now, $z' \in T$, so we can find $z \leq_\lambda z'$, $z \in D$. We may assume $z \in T_{\delta+1}$, where $\delta > \gamma$. Thus $z \in T^{\lambda(\delta+1)}_{\beta_{\delta+1}}$. Let $y'' = h_{\lambda(\delta+1),\lambda(\nu)}(z)$. Then $y'' \leq_{\lambda(\nu)} y$ and $y'' \in E_\nu$.

<u>Claim 6</u>. There is a closed unbounded set $K_2 \subseteq \omega_1$ such that for all $\alpha \in K_2$, all $\nu < \alpha$, and all $y \in T^{\lambda(\nu)}_\alpha$, $(\exists \tau < \alpha)(\exists z \in T^{\lambda(\tau)}|\alpha)[\tau \geq \nu$.&. $z \in D_\alpha$.&. $h_{\lambda(\tau),\lambda(\nu)}(z) \geq y]$.

For each $\nu < \omega_1$, let Y_ν be a maximal disjointed subset of E_ν. Define maps φ, ψ on ω_1 by:

$\varphi(\nu)$ = [the least $\varphi < \omega_1$ such that $Y_\nu \subseteq T^{\lambda(\nu)}|\varphi$] and

$\psi(\nu)$ = [the least $\psi < \omega_1$ such that for all $y \in Y_\nu$, $(\exists\tau < \psi)$ $(\exists z \in T^{\lambda(\tau)}|\psi)[\tau \geq \nu$.&. $z \in D_\psi$.&. $h_{\lambda(\tau),\lambda(\nu)}(z) = y]$. Let $K_2 = \{\alpha \in \omega_1 \mid (\forall\nu < \alpha)(\varphi(\nu) < \alpha$.&. $\psi(\nu) < \alpha)\}$. We show that K_2 is as required. Let $\alpha \in K_2$, $\nu < \alpha$, $y \in T_\alpha^{\lambda(\nu)}$. By claim 5, $\bigvee Y_\nu = 1\!\!1$, so we can find $y' \in Y_\nu$, $y \leq_{\lambda(\nu)} y'$. For some $\tau < \alpha$, $\tau \geq \nu$, and some $z \in T^{\lambda(\tau)}|\alpha$, $z \in D_\alpha$, we have $y' = h_{\lambda(\tau),\lambda(\nu)}(z)$. Thus K_2 is as claimed.

Let $K = K_0 \cap K_1 \cap K_2$. Let $D^* = \{\langle\tau,z\rangle \mid \tau < \omega_1$.&. $z \in \tilde{T}^{\lambda(\tau)}$.&. $\ell^{-1}_{\lambda(\tau)}(z) \in D\}$, and for $\alpha < \omega_1$, set $D^*_\alpha = \{\langle\tau,z\rangle \mid \tau < \alpha$.&. $z \in \tilde{T}^{\lambda(\tau)}|\alpha$.&. $\ell^{-1}_{\lambda(\tau)}(z) \in D_\alpha\}$. Clearly $\alpha \in K \rightarrow D^*_\alpha = D^* \cap V_\alpha$. By $\Diamond$, therefore, we can find $\alpha \in K$ such that $S_\alpha = D^*_\alpha$. Consider such an α. We recall the construction of $\mathbb{B}_{\lambda(\alpha)}$. Notice first that $A_{\lambda(\alpha)} = A_\lambda \cap \lambda(\alpha)$ and that otp $(A_{\lambda(\alpha)}) = \alpha$. To obtain $\mathbb{B}_{\lambda(\alpha)}$, we defined a Souslin tree $\langle\hat{T},\leq^*\rangle$ whose elements consisted of sequences $x = \langle x_\nu \mid \nu < \alpha\rangle$ such that for some $\gamma \in C$, $\nu < \alpha \rightarrow x_\nu \in \tilde{T}_\gamma^{\lambda(\nu)}$ and $\nu \leq \tau < \alpha \rightarrow x_\nu = \tilde{h}_{\lambda(\tau),\lambda(\nu)}(x_\tau)$, where $C = [\bigcap_{\nu<\tau<\alpha}(C_{\lambda(\nu),\lambda(\tau)} - \alpha)] \cup \{0\}$. Let $\langle c_\nu | \nu < \omega_1\rangle$ enumerate C. Since $\alpha = \beta_\alpha \in B_\alpha = \bigcap_{\nu<\tau<\alpha} C_{\lambda(\nu),\lambda(\tau)}$, $\alpha \in C$. Thus, by definition of C, $\alpha = c_1$. But look, $\alpha \in K$ and $S_\alpha = D^*_\alpha$, so by choice of $K_2 \subseteq K$ we see that the special case (*) must have applied in defining $\hat{T}_1$. Thus, by construction, for every $x \in \hat{T}_1$ there is $\langle\tau,z\rangle \in D^*_\alpha$ such that $x_\tau \leq_{\lambda(\tau)} z$. Define $\eta : \hat{T} \rightarrow \mathbb{B}_{\lambda(\alpha)}$ by $\eta(x) = \bigwedge_{\nu<\alpha} \ell^{-1}_{\lambda(\nu)}(x_\nu)$ as before. Recalling that $\eta''\hat{T}$ is a Souslinisation of $\mathbb{B}_{\lambda(\alpha)}$, we see that, in particular, $x \in \hat{T}_1 \rightarrow \eta(x) > 0$, and that $\bigvee^{\mathbb{B}_{\lambda(\alpha)}} \eta''\hat{T}_1 = 1\!\!1$. So, as $\alpha = \beta_\alpha = c_1$, the definition of T_α implies that $T_\alpha = \eta''\hat{T}_1$. Let $y \in T_\alpha$, ie. $y \in \eta''\hat{T}_1$. By our earlier observations, there are $\nu \leq \tau < \alpha$ and $z \in D_\alpha \cap (T^{\lambda(\tau)}|\alpha)$ such that $h_{\lambda(\alpha),\lambda(\nu)}(y) \leq_{\lambda(\nu)} h_{\lambda(\tau),\lambda(\nu)}(z)$. Thus, $y \leq_{\lambda(\alpha)} h_{\lambda(\tau),\lambda(\nu)}(z)$, which implies $y \wedge z >_\lambda \mathbb{0}$. But T is a tree and $y, z \in T$. Hence $y \leq_\lambda z$. Hence , every $y \in T_\alpha$ extends an element of D_α . In other words, every $y \in T_\alpha$ extends an element of X_α, whence X_α is a maximal antichain of $T|\alpha$ and of T , proving that $X = X_\alpha$, which is countable.

It remains to prove that $\nu < \lambda \rightarrow \mathbb{B}_\nu$ is a nice subalgebra of $\mathbb{B}_\lambda$. By lemma 11, it suffices to prove $\nu < \lambda \rightarrow \mathbb{B}_{\lambda(\nu+1)}$ is a nice subalgebra of $\mathbb{B}_\lambda$. Let $\nu < \omega_1$, therefore, and let $h_\nu : \mathbb{B}_\lambda \rightarrow \mathbb{B}_{\lambda(\nu+1)}$ be the basic projection. For all $\alpha = \beta_\alpha > \nu$, we clearly have $T_\alpha^{\lambda(\nu+1)} = h_\nu{}''T_\alpha$. QED. ▮

Chapter IX

HOW JENSEN KILLED A SOUSLIN TREE

In this chapter we shall describe how to kill a Souslin tree in a manner which can be fitted into the iteration apparatus of the last chapter. As we have already hinted, our method will be to embed the given Aronszajn tree into $\mathbb{Q}$. However, the way in which we shall go about this is by no means straightforward, and the reader must bear with us if we appear to be losing sight of our goal, as we shall from the very start, in fact, appear to so do.

We commence by describing a certain method for generically adding a closed unbounded subset of ω_1 to a model of ZFC. Define a poset $\mathbb{C} = \langle C, \leq \rangle$ by setting $\mathbb{C} = \{\langle \nu, A \rangle \mid \nu < \omega_1$ & A is a closed unbounded subset of $\omega_1\}$ and putting $\langle \nu', A' \rangle \leq \langle \nu, A \rangle$ iff $\nu' \geq \nu$ & $A' \subseteq A$ & $\nu \cap A' = \nu \cap A$. Clearly, $\mathbb{C}$ is σ-closed and satisfies ω_2-c.c. We call $BA(\mathbb{C})$ the <u>closed set algebra</u>.

Let M be a c.t.m. of ZFC and consider $\mathbb{C}^M$. Let G be M-generic on $\mathbb{C}^M$. Set $C_G = \bigcup\{\nu \cap A \mid \langle \nu, A \rangle \in G\}$. Clearly, C_G is a closed unbounded subset of ω_1 in $M[G]$. In fact, $C_G = \bigcap\{A \mid \langle 0, A \rangle \in G\}$, as is easily verfied. Since $G = \{\langle \nu, A \rangle \mid \nu \cap A \subseteq C_G \subseteq A\}$, $M[C_G] = M[G]$, so we call any such C_G an <u>M-generic subset of ω_1^M over $\mathbb{C}^M$</u>.

<u>Lemma 1</u>

Let C be an M-generic subset of ω_1^M over $\mathbb{C}^M$. Let $A \in M$ be a closed unbounded subset of ω_1^M (in M). Then there is $\alpha < \omega_1^M$ such that $C - \alpha \subseteq A$. Again, if $\langle \gamma_\nu \mid \nu < \omega_1^M \rangle$ is the normal enumeration of C (in $M[C]$), and if $f \in M$ is a normal function on ω_1^M, then there is $\alpha < \omega_1^M$ such that $\gamma_\alpha = \alpha$ and $(\forall \beta \geq \alpha)(f(\gamma_\beta) = \gamma_\beta)$.

Proof: Trivial. ▮

In M, define $\mathring{C} : \mathrm{dom}(\check{\omega}_1) \to BA(\mathbb{C})$ by $\mathring{C}(\check{\tau}) = \{\langle \nu, A\rangle \mid \tau \in \nu \cap A\}$ (where the "hats" are calculated for $BA(\mathbb{C})$ in M). Clearly, if C is M-generic over $\mathbb{C}^M$, then $C = \mathring{C}^{M[C]}$. Also, $\Vdash_{\mathbb{C}}$ "$\mathring{C} \subseteq \check{\omega}_1$" and $\langle \nu, A\rangle \Vdash_{\mathbb{C}}$ "$\check{\tau} \in \mathring{C}$" iff $\tau \in \nu \cap A$.

It will be convenient for us to redefine the notion of genericity. Suppose U is a transitive set (but not necessarily a model of ZF). Let $\mathbb{P}$ be a poset in U. We shall call a set $G \subseteq \mathbb{P}$ <u>U-generic for $\mathbb{P}$</u> iff G is a pairwise compatible final section of $\mathbb{P}$ which meets every dense subset of $\mathbb{P}$ <u>which is first-order definable in</u> $\langle U, \in, (x)_{x\in U}\rangle$. Using this definition, the forcing relation for $U, \mathbb{P}$ is definable in U, and is in fact primitive recursive for Σ_0 statements. We can, of course, assume that this definition of genericity is the one which is used when U is a ZF-model and we are forming a generic extension of U, since in that case the two definitions clearly coincide.

Suppose $N \prec H_{\omega_2}$, $|N| = \omega$. Let $\pi_N : N \cong \bar{N}$, where $\bar{N}$ is transitive. Set $\alpha_N = \omega_1 \cap N$. Then, $\alpha_N \in \omega_1$, $\pi_N(\omega_1) = \alpha_N$, $\pi_N(H_{\omega_1}) = H_{\omega_1} \cap N$, $x \in H_{\omega_1} \cap N \to \pi_N(x) = x$, $x \in \mathcal{P}(H_{\omega_1}) \cap N \to \pi_N(x) = x \cap N$, $\langle \nu, A\rangle \in \mathbb{C} \cap N \to \pi_N(\langle \nu, A\rangle) = \langle \nu, A \cap \alpha_N\rangle$. We use $\pi_N(\mathbb{C})$ to denote $\pi_N``(\mathbb{C} \cap N)$.

<u>Lemma 2</u>

Let $N \prec H_{\omega_2}$, $|N| = \omega$, $p \in \mathbb{C} \cap N$. Let U be any countable transitive set with $\bar{N}, \pi_N(\mathbb{C}) \in U$. Then there is $p' \leq p$ of the form $p' = \langle \alpha_N, C\rangle$ such that:

(i) $C \cap \alpha_N$ is a U-generic subset of α_N over $\pi_N(\mathbb{C})$;

(ii) if φ is any sentence of $V^{(BA(\mathbb{C}))}$ which involves only constants from $\{\check{x} \mid x \in N\} \cup \{\mathring{C}\}$, then $\bar{N}[C \cap \alpha_N] \models \pi_N(\varphi)$ iff

$p' \Vdash_{\mathbb{C}} "H_{\omega_2} \models \varphi"$.

Proof: Let $p = \langle \nu, A \rangle$. Let G be U-generic on $\pi_N(\mathbb{C})$ with $\pi_N(p) \in G$. Set $C = \bigcap\{\pi_N^{-1}(B) \mid \langle 0,B\rangle \in G\}$. Now, if $\langle 0,B\rangle \in G$, B is closed and unbounded in α_N , so $\pi_N^{-1}(B)$ is closed and unbounded in ω_1 . Hence $p' = \langle \alpha_N, C\rangle \in \mathbb{C}$. Since $\pi_N(p) = \langle \nu, A \cap \alpha_N\rangle \in G$, $\langle 0, A \cap \alpha_N\rangle \in G$, whence $C \subseteq \pi_N^{-1}(A \cap \alpha_N) = A$. But look, since $\langle \nu, A \cap \alpha_N\rangle \in G$, we also have $C \cap \nu = A \cap \nu$. Thus $p' \leq p$. Hence, by choice of G , (i) holds. Notice also that for any $q \in G$, $p' \leq_{\mathbb{C}} \pi_N^{-1}(q)$. We now verify (ii). By (i), $\bar{N}[C \cap \alpha_N] \models \pi_N(\varphi)$ iff $(\exists q \in G)(q \Vdash_{\pi_N(\mathbb{C})} \pi_N(\varphi))$. Now, $\pi_N^{-1} : \bar{N} \prec H_{\omega_2}$, so if $\bar{N}[C \cap \alpha_N] \models \pi_N(\varphi)$, then $(\exists q \in G)(\pi_N^{-1}(q) \Vdash_{\mathbb{C}} \varphi)$, which implies that $p' \Vdash_{\mathbb{C}} \varphi$, so by choice of φ , $p' \Vdash_{\mathbb{C}} "H_{\omega_2} \models \varphi"$. Similarly for $\neg\varphi$, proving (ii). ∎

Lemma 3

Let M be a c.t.m. of ZFC. In M , let U be a function such that for each countable $N \prec H_{\omega_2}$, U_N is a countable transitive set with $\bar{N}$, $\pi_N(\mathbb{C}^M) \in U_N$. Let C be an M-generic subset of ω_1^M over $\mathbb{C}^M$. Let $x \in H_{\omega_2}^M$. Then there is a countable $N \prec H_{\omega_2}$ in M such that $x \in N$ and such that, in $M[C]$, we have: -

(i) $C \cap \alpha_N$ is a U_N-generic subset of α_N over $\pi_N(\mathbb{C}^M)$; and

(ii) there is a map $\pi \supseteq \pi_N$ such that $\pi(C) = C \cap \alpha_N$ and $\pi^{-1} : \bar{N}[C \cap \alpha_N] \prec H_{\omega_2}$.

Proof: By lemma 2 and a simple density argument for $\mathbb{C}^M$ forcing over M . ∎

So much (for now) for the closed set algebra. It would seem to be in order for us to state, precisely, the result which we shall spend the rest of this chapter in proving.

CRUCIAL LEMMA

Let M be a c.t.m. of ZFC + $(2^{\omega} = \omega_1) + (2^{\omega_1} = \omega_2) + \Diamond^*$. Let $\mathbb{B}$ be a Souslin algebra in M, and let $T_{\mathbb{B}}$ be, in M, an arbitrary Souslinisation of $\mathbb{B}$. Let $\mathring{T} \in M^{(\mathbb{B})}$ be, with $\mathbb{B}$-value $\mathbb{1}$, an Aronszajn tree. Let C be an M-generic subset of ω_1^M over $\mathbb{C}^M$. Then there is a Souslin algebra $\mathbb{B}'$ in M[C] such that, in M[C], $\mathbb{B}$ is a nice subalgebra of $\mathbb{B}'$ and $\|\mathring{T}$ is special$\|^{\mathbb{B}'} = \mathbb{1}$.

Note that we do not find our Souslin algebra $\mathbb{B}'$ in M, but must first of all pass to M[C]. That this will not cause any problems when we try to fit the crucial lemma into our iteration scheme, will become clear in the next chapter.

From now on we fix M, $\mathbb{B}$, $T_{\mathbb{B}}$, $\mathring{T}$, C as above. We may assume that $\|\mathring{T}$ has domain $\check{\omega}_1\|^{\mathbb{B}} = \mathbb{1}$. It will not cause any ambiguity if we write T for $T_{\mathbb{B}}$, since when we come to consider the evaluation of $\mathring{T}$ in a generic extension over $\mathbb{B}$ we shall introduce a special notation for it. (So remember, T and $\mathring{T}$ are quite different animals!) We shall also assume that $\alpha \in C \rightarrow \lim(\alpha)$. (This will clearly not cause any headaches, either.)

Note that, in M, $\|T$ has an ω_1-branch$\|^{\mathbb{B}} = \mathbb{1}$. In fact by II.7 we see that any cofinal branch of T (in V) will be M-generic on $\mathbb{B}$ and vice-versa. Bearing this in mind, we prove:

Lemma 4

Let b be a cofinal branch of T, and let $A \in M[b]$ be a closed unbounded subset of ω_1. There is $B \in M$, $B \subseteq A$, B closed and unbounded in ω_1.

Proof: In M[b], let $\langle a(\nu) \mid \nu < \omega_1 \rangle$ enumerate A. Let $\mathring{a} \in M^{(\mathbb{B})}$ be such that $\|\mathring{a}$ is a normal function on $\check{\omega}_1\|^{\mathbb{B}} = \mathbb{1}$ and

$\mathring{a}^{M[b]} = a$.

We say that an element p of T <u>fixes</u> $\mathring{a}(\check{\nu})$ iff $p \Vdash_{\mathbb{B}}$ "$\mathring{a}(\check{\nu}) = \check{\tau}$" for some $\tau < \omega_1$. Work in M . For each $\nu < \omega_1$, let X_ν be a maximal disjoint set such that $X_\nu \subseteq T$ and $p \in X_\nu \rightarrow p$ fixes $\mathring{a}(\check{\nu})$. For each ν , let $\beta(\nu)$ be least such that $X_\nu \subseteq T|\beta(\nu)$. Then, for each ν , if $p \in T_{\beta(\nu)}$, p fixes $\mathring{a}(\check{\nu})$. Let $B = \{\alpha < \omega_1 \mid (\forall \nu < \alpha)(\forall p \in T_{\beta(\nu)}(\text{if } p \Vdash \text{"}\mathring{a}(\check{\nu}) = \check{\tau}\text{", then } \tau < \alpha)\}$. Clearly, every element of the closed unbounded set B is a limit point of A : ∎

<u>Lemma 5</u>

There is a sequence $\langle W_\alpha \mid \alpha < \omega_1 \rangle$ in M such that:

(i) $M \models (\forall \alpha < \omega_1)(W_\alpha \subseteq \mathcal{P}(\alpha)\ \&\ |W_\alpha| \leq \omega)$;

(ii) if b is a cofinal branch of T and $A \in M[b]$, $A \subseteq \omega_1$, then $\{\alpha \in \omega_1 \mid A \cap \alpha \in W_\alpha\}$ contains a closed unbounded set $B \in M$.

Proof: Let b be a cofinal branch of T . Then, by VIII.4 $M[b] \models \diamondsuit^*$. Hence, we can find $\mathring{W} \in M^{(\mathbb{B})}$ such that, in M , $\|\mathring{W}$ satisfies $\diamondsuit^*\|^{\mathbb{B}} = 1\!\!1$. Work in M . Since $\mathbb{B}$ is ω_1-dense, we can extend every $p \in T$ to a $q \in T$ which <u>fixes</u> $\mathring{W}_{\check{\alpha}}$ (i.e. $q \Vdash$ "$\mathring{W}_{\check{\alpha}} = \check{w}$", for some $w \in M$), each $\alpha < \omega_1$. Then, as above, for each $\alpha < \omega_1$ we can find $\beta = \beta(\alpha) < \omega_1$ such that $p \in T_\beta \rightarrow p$ fixes $\mathring{W}_{\check{\alpha}}$. For each α , each $p \in T_{\beta(\alpha)}$, set $W(\alpha,p)$ = that $w \in M$ such that $p \Vdash$ "$\mathring{W}_{\check{\alpha}} = \check{w}$". Set $W_\alpha = \bigcup\{W(\alpha,p) \mid p \in T_{\beta(\alpha)}\}$, $\alpha < \omega_1$. Then, using lemma 4, $\langle W_\alpha \mid \alpha < \omega_1 \rangle$ is easily seen to be as required. ∎

<u>Lemma 6</u>

There is a sequence $\langle W_\alpha \mid \alpha < \omega_1 \rangle$ in M such that:

(i) $M \models (\forall \alpha < \omega_1)(W_\alpha \subseteq H_{\omega_1}(\alpha+1)\ \&\ |W_\alpha| \leq \omega)$;

(ii) if b is a cofinal branch of T and $A \in M[b]$, $A \subseteq H_{\omega_1}$, and

$(\forall\alpha<\omega_1)(|A\cap H_{\omega_1}(\alpha)|\leq\omega)$, then $\{\alpha\in\omega_1 \mid A\cap V_\alpha\in W_\alpha\}$ contains a closed unbounded set $B\in M$.

Proof: This follows from lemma 5 by an argument as in lemma VIII.5. ∎

Fix $\langle W_\alpha \mid \alpha<\omega_1\rangle$ as in lemma 6 for the rest of this proof.

We say that a $p\in T$ <u>fixes</u> $\mathring{T}|\check{\alpha}$ iff there is a normal α-tree t in M such that $p \Vdash "\mathring{T}|\check{\alpha}=\check{t}"$.

As before, for each $\alpha<\omega_1$ there is a least ordinal $\beta(\alpha)<\omega_1$ such that every $p\in T_{\beta(\alpha)}$ fixes $\mathring{T}|\check{\alpha}$. Let $\langle\eta(\alpha) \mid \alpha<\omega_1\rangle$ enumerate the closed unbounded set $\{\eta\in\omega_1 \mid \nu<\eta \rightarrow \beta(\nu)<\eta\}$. Thus, for all $\alpha<\omega_1$, $p\in T_{\eta(\alpha)} \rightarrow p$ fixes $\mathring{T}|\widetilde{\eta(\alpha)}$. For such α, p , let $\mathring{T}_p$ be that normal $\eta(\alpha)$-tree t in M such that $p \Vdash "\mathring{T}|\widetilde{\eta(\alpha)}=\check{t}"$. Since $\mathbb{1} \Vdash \mathring{T}|\check{0}=\emptyset$, we may assume $\eta(0)=0$ in all of this. Let $\mathcal{A}_\eta = \eta''\omega_1$.

Let $\langle\gamma_\alpha \mid \alpha<\omega_1\rangle$ be the normal enumeration of C . By lemma 1, we can find an $\alpha<\omega_1$ such that $\gamma_\alpha=\alpha$ and $(\forall\beta\geq\alpha)(\eta(\gamma_\beta)=\gamma_\beta)$. We may assume that $\alpha=0$ here. (This is because $C-\gamma_\alpha$ differs from C only on an initial segment, so will itself be M-generic over $\mathbb{C}^M$, which is all that we need to know about C !)

Note that if b is a cofinal branch of T , then $\mathring{T}^{M[b]} = \bigcup\{\mathring{T}_p \mid p\in b \;\&\; ht_T(p)\in C\}$.

We wish to construct a Souslin algebra $\tilde{\mathbb{B}}$ in $M[C]$ such that $\mathbb{B}$ is a nice subalgebra of $\tilde{\mathbb{B}}$ and, in $M[C]^{(\tilde{\mathbb{B}})}$ it holds with $\tilde{\mathbb{B}}$-value $\mathbb{1}$ that $\mathring{T}$ is special.

Our strategy is to construct a Souslin tree $\tilde{T}$ in $M[C]$ with the following properties:

(i) There is a map $h : \tilde{T} \to T$ such that:

(a) $x <_{\tilde{T}} y \to h(x) <_T h(y)$;

(b) $h : \tilde{T}_\nu \xrightarrow{\text{ONTO}} T_{\gamma_\nu}$, all $\nu < \omega_1$;

(c) if $z <_T h(x)$ and $ht_T(z) \in C$, there is $y \in \tilde{T}$, $y <_{\tilde{T}} x$, such that $h(y) = z$.

(Thus h resembles a neat cover very strongly.)

Set $\tilde{\mathbb{B}} = BA(\tilde{T})$. [Note that, by (i), there is $\tilde{h} : \mathbb{B} \to \tilde{\mathbb{B}}$ which nicely embeds $\mathbb{B}$ in $\tilde{\mathbb{B}}$. In fact, the restriction of $\tilde{h}$ to $\bigcup_{\nu<\omega_1} T_{\gamma_\nu}$ is defined by $\tilde{h}(z) = \{x \in \tilde{T} \mid (\exists y \in \tilde{T})(x \leq_{\tilde{T}} y \ \& \ h(y) = z)\}$. Thus, when we are done, all we shall need to do is replace $\tilde{\mathbb{B}}$ by an isomorph around $\mathbb{B}$ to make $\tilde{h}$ the identity. We shall not bother ourselves with this point again.]

(ii) $\|\mathring{\tilde{T}}$ is special$\|^{\tilde{\mathbb{B}}} = \mathbb{1}$.

(iii) The elements of $\tilde{T}$ are pairs $\langle x,f \rangle$ such that $x \in T$, $ht_T(x) \in C$, and f is an order-preserving map of $\mathring{T}_x | C$ ($= \bigcup\{(\mathring{T}_x)_\alpha \mid \alpha < ht_T(x) \ \& \ \alpha \in C\}$) into $\mathbb{Q}$ (the rationals).

(Remark: Since our trees "grow downwards", we shall assume the rationals "increase" likewise.)

(iv) $\langle x,f \rangle \leq_{\tilde{T}} \langle x',f' \rangle \leftrightarrow x \leq_T x' \ \& \ f \supseteq f'$.

(v) $h(\langle x,f \rangle) = x$.

In order that our definition does not break down, we ensure that at every stage the following two conditions hold:

(*) If $\lim(\alpha)$ and $\langle x,f \rangle \in \tilde{T}_{\alpha+1}$, then for all $t \in (\mathring{T}_x)_{\gamma_\alpha}$, $f(t) = \sup\{f(s) \mid s >_{\mathring{T}_x} t\}$.

(**) Let $\langle x,f \rangle \in \tilde{T}_{\alpha+1}$, $x' <_T x$, $ht(x') \in C$. Let $b_1,\ldots,b_n$ be cofinal branches of $\mathring{T}_x$ which extend to points $t_1,\ldots,t_n$ in $(\mathring{T}_{x'})_{ht_T(x)}$. Let $q_1,\ldots,q_n \in Q$ be such that $q_i > \sup(f''[b_i \upharpoonright C])$, $i = 1,\ldots,n$. (Note that each $b_i \upharpoonright C$ will have a maximal element.) Then there is an f' such $\langle x',f'\rangle \leq_{\tilde{T}} \langle x,f\rangle$ and $f'(t_i) = q_i$, $i = 1,\ldots,n$.

As usual, ZF^- denotes ZF minus the power set axiom.

We define sequences $\langle \delta(\nu) \mid \nu < \omega_1\rangle$ and $\langle N_\nu \mid \nu < \omega_1\rangle$ by a simultaneous recursion, as follows.

Let $\delta(\alpha)$ be the least $\delta > \alpha$ such that $L_\delta[\langle N_\nu \mid \nu < \alpha\rangle, W_\alpha, C \cap \alpha]$ is a ZF^--model, and for all $b \in W_\alpha$:

(i) if α is countable in $L[b, C \cap \alpha]$, then α is countable in $L_\delta[b, C \cap \alpha]$;

(ii) if α is uncountable in $L[b, C\cap\alpha]$ but $(\alpha^+)^{L[b,C\cap\alpha]} < \omega_1$, then $\delta > (\alpha^+)^{L[b,C\cap\alpha]}$.

Let $N_\alpha = L_{\delta(\alpha)}[\langle N_\nu \mid \nu < \alpha\rangle, W_\alpha, C \cap \alpha]$.

The definition of $\tilde{T}$ is by induction on the levels. As we proceed, we set

$$F_\alpha = \langle \mathring{T}_y \mid y \in T|C \cap \gamma_{\alpha+2}\rangle ;$$
$$\tilde{N}_\alpha = N_{\gamma_{\alpha+2}}[\tilde{T}|\alpha, T|\gamma_{\alpha+2}, F_\alpha] .$$

Set $\tilde{T}_0 = \{\langle \mathbb{1}_{\mathbb{B}}, \emptyset\rangle\}$;
and $\tilde{T}_1 = \{\langle x, \langle 0,0\rangle\rangle \mid x \in T_{\gamma_1}\}$ (where we have assumed that $\|\mathring{T}_0 = \{0\}\|^{\mathbb{B}} = \mathbb{1}$ for convenience).

Suppose $\tilde{T}_{\alpha+1}$ is defined and $\tilde{T}|(\alpha+2)$ satisfies (*) and (**) . We define $\tilde{T}_{\alpha+2}$ by forcing over $\tilde{N}_{\alpha+2}$.

Let $S = \{\langle x,\langle x',f\rangle\rangle \mid x \in T_{\gamma_{\alpha+2}}$ & $x <_T x' \in T_{\gamma_{\alpha+1}}$ & $\langle x',f\rangle \in \tilde{T}|(\alpha+2)\}$.

For each $s = \langle x,\langle x',f\rangle\rangle \in S$, let $\mathbb{P}^s = \{p \mid p$ is a finite function & $\mathrm{dom}(p) \subseteq (T_x^\circ)_{\gamma_{\alpha+1}}$ & $\mathrm{ran}(p) \subseteq \mathbb{Q}$ & $(\forall t \in \mathrm{dom}(p))(\forall s \in (T_{x'}^\circ)_{\gamma_\alpha})(t <_{T_1^\circ} s \rightarrow p(t) >_{\mathbb{Q}} f(s))\}$. Regard $\mathbb{P}^s$ as a poset under $\supseteq$. Note that $\mathbb{P}^s \in \tilde{N}_{\alpha+2}$.

For each $s \in S$ as above and each $p \in \mathbb{P}^s$, we define an element of $\tilde{T}_{\alpha+2}$ as follows.

Let $X_{s,p}$ be an $\tilde{N}_{\alpha+2}$-generic subset of $\mathbb{P}^s$ with $p \in X_{s,p}$. Put $\langle x, f \cup X_{s,p}\rangle$ into $\tilde{T}_{\alpha+2}$. Note that $\langle x,f \cup X_{s,p}\rangle \leq_{\tilde{T}} \langle x',f\rangle$.

It is obvious that $\tilde{T}|(\alpha+3)$ still satisfies (*) and (**).

Suppose next that $\lim(\alpha)$, $\alpha < \omega_1$, and $\tilde{T}|\alpha$ is defined and satisfies (*), (**). We construct $\tilde{T}_\alpha$ by forcing over $\tilde{N}_\alpha$.

For each $x \in T_{\gamma_\alpha}$, let $B_x = \{b \mid b$ is a cofinal branch of T_x° such that, for some $y \in T_{\gamma_{\alpha+1}}$ with $y <_T x$, there is $z \in (T_y^\circ)_{\gamma_\alpha}$ which extends $b\}$.

For each $x \in T_{\gamma_\alpha}$, let $\mathbb{P}^x = \{\langle\langle x',f'\rangle,p\rangle \mid \langle x',f'\rangle \in \tilde{T}_{\beta+1}$ for some $\beta < \alpha$ & $x <_T x'$ & p is a finite function & $\mathrm{dom}(p) \subseteq B_x$ & $\mathrm{ran}(p) \subseteq \mathbb{Q}$ & $(\forall b \in \mathrm{dom}(p))(\forall t \in T_{x'}^\circ|C)(t \in b \rightarrow f'(t) <_{\mathbb{Q}} p(b))\}$. Regard $\mathbb{P}^x$ as a poset under $\langle\langle x',f'\rangle,p\rangle \leq \langle x'',f''\rangle,p'\rangle \leftrightarrow \langle x',f'\rangle \leq_{\tilde{T}} \langle x'',f''\rangle$ & $p \supseteq p'$. Note that $\mathbb{P}^x \in \tilde{N}_\alpha$.

For each $x \in T_{\gamma_\alpha}$ and each $u \in \mathbb{P}^x$, we define an element of $\tilde{T}_\alpha$ as follows. Let $X_{x,u}$ be an $\tilde{N}_\alpha$-generic subset of $\mathbb{P}^x$ such that $u \in X_{x,u}$. Set $f_{x,u} = \bigcup\{f' \mid \langle\langle x',f'\rangle,\emptyset\rangle \in X_{x,u}\}$. Since (**) holds for $\tilde{T}|\alpha$ (and to some extent since (*) holds also), it is easily seen that

$f_{x,u}$ is an order-preserving map of $\mathring{T}_x|C$ into $\mathbb{Q}$. Set $g_{x,u} = \bigcup\{p \mid (\exists d)(\langle d,p\rangle \in X_{x,u})\}$. Clearly, $g_{x,u} : B_x \to \mathbb{Q}$ and for each $b \in B_x$, $g_{x,u}(b) = \sup(f_{x,u}"[b\restriction C])$. Put $\langle x,f_{x,u}\rangle$ into $\tilde{T}_\alpha$. Note that $\langle x,f_{x,u}\rangle \leq_{\tilde{T}} \langle x',f'\rangle$, where $u = \langle\langle x',f'\rangle,p\rangle$.

There are no new cases of (*), (**) to consider here.

Finally, suppose $\lim(\alpha)$, $\alpha < \omega_1$, and $\tilde{T}|(\alpha+1)$ is defined (as above). We define $\tilde{T}_{\alpha+1}$ to make (*) hold.

For each $x \in T_{\gamma_\alpha}$, $y \in T_{\gamma_{\alpha+1}}$, $y <_T x$, and each $u \in \mathbb{P}^x$, define $\bar{f}_{y,x,u}$ from $\mathring{T}_y|C$ into $\mathbb{Q}$ by

$$\bar{f}_{y,x,u}(t) = \begin{cases} f_{x,u}(t) , & \text{if } t \in \mathring{T}_x \\ g_{x,u}(b) , & \text{if } t \in (\mathring{T}_y)_{\gamma_\alpha} \ \& \ b = \{s \in \mathring{T}_x \mid t <_{\mathring{T}_y} s\}. \end{cases}$$

Noting that, in the above, x is uniquely determined by y , we obtain $\tilde{T}_{\alpha+1}$ by taking $\langle y,\bar{f}_{y,x,u}\rangle$ for each pair $\langle y,u\rangle$ as above. Note that $\langle y,\bar{f}_{y,x,u}\rangle \leq_{\tilde{T}} \langle x,f_{x,u}\rangle$.

Clearly (*) holds for $\tilde{T}|(\alpha+2)$, whilst no <u>essentially</u> new case of (**) arises.

Set $\tilde{T} = \bigcup_{\alpha<\omega_1}\tilde{T}_\alpha$, a normal ω_1-tree. It is clear that (i),(iii)-(v) hold. We verify (ii).

Let b be $M[C]$-generic on $\tilde{T}$. Set $b_0 = \{x \in T \mid (\exists y \in T)(y \leq_T x \ \& \ \langle y,f\rangle \in b)\}$. By lemma VII.4, T is a Souslin tree in $M[C]$, so b_0 is $M[C]$-generic on T . Let $\mathring{T}_b = \bigcup\{\mathring{T}_x \mid x \in b_0\restriction C\}$. Clearly, $\mathring{T}_b = \mathring{T}^{M[C][b]}$. Set $f_b = \bigcup\{f \mid \langle x,f\rangle \in b\}$. Then f_b is an orderpreserving map from $\mathring{T}^{M[C][b]}|C$ into $\mathbb{Q}$ in $M[C][b]$. But look, C is a closed unbounded subset of ω_1 in $M[C][b]$, so, using f_b , we can easily define, in $M[C][b]$, an order-embedding of all of T_b° into $\mathbb{Q}$. Hence, $M[C][b] \models$ "$\mathring{T}$ is special" , whence, by choice of b , $M[C] \models \| \mathring{T}$ is special$\|^{\mathbb{B}} = \mathbb{1}$.

It remains to check that $\tilde{T}$ is Souslin, We require two lemmas, the second of which may appear rather strange at first glance.

Lemma 7

Let $D \in M[C]$ be dense in $\tilde{T}$. Let b be a cofinal branch of T . Set $\tilde{T}_b = \{\langle x,f\rangle \in \tilde{T} \mid x \in b\}$. Then $D \cap \tilde{T}_b$ is dense in $\tilde{T}_b$.

Proof: Since T is Souslin in $M[C]$, b is $M[C]$-generic on T . Thus, if $D \cap \tilde{T}_b$ is not dense in $\tilde{T}_b$ we can find $x_o \in b$ and $\langle x,f\rangle \in \tilde{T}_b$ such that $x_o \Vdash_{\mathbb{B}}$"$\langle x,f\rangle \in \check{\tilde{T}}_{\mathring{b}}$ & $\langle x,f\rangle$ has no extension in $\check{D}$" , where the forcing here is over $M[C]$, and where $\mathring{b}$ is the generic set term for $M[C]$, $\mathbb{B}$ forcing. We may clearly assume $x = x_o$ here. Pick $\langle x',f'\rangle \in D$, $\langle x',f'\rangle \leq_{\tilde{T}} \langle x,f\rangle$. Then $x' \Vdash_{\mathbb{B}}$ "$\langle x',f'\rangle \in \check{\tilde{T}}_{\mathring{b}} \cap \check{D}$" , contrary to $x' \leq_T x_o$.

Lemma 8

Let b be a cofinal branch of T , and set $M' = M[b]$, $T^\circ = \bigcup\{T_x^\circ \mid x \in b \restriction \not{\!\lambda}_\eta\}$. Let $b_1,\dots,b_n$ be cofinal branches of T° , and suppose that C is an $M'[b_1,\dots,b_n]$-generic subset of ω_1^M over $\mathbb{C}^M$, with $\tilde{T}$ defined in $M[C]$ as above, still. Set $\tilde{T}_b = \{\langle x,f\rangle \in \tilde{T} \mid x \in b\}$. Let $p:\{b_1,\dots,b_n\} \to \mathbb{Q}$, and set $\tilde{T}_b^p = \{\langle x,f\rangle \in \tilde{T}_b \mid \bigwedge_{i=1}^{n} (\exists q_i <_{\mathbb{Q}} p(b_i))(\forall t \in T_x^\circ | C)(t \in b_i \to f(t) <_{\mathbb{Q}} q_i)\}$. Let $D \in M'[C]$ be dense in $\tilde{T}_b$. Then $D^\bullet \cap \tilde{T}_b^p$ is dense in $\tilde{T}_b^p$.

Proof: Let $p(b_i) = r_i$, $i = 1,\dots,n$. For each $\langle x,f\rangle \in \tilde{T}_b$, let $[\![\langle x,f\rangle]\!]_i = \{z \in T^\circ \mid ht_{T^\circ}(z) = ht_T(x)$ & $(\exists r_i' <_{\mathbb{Q}} r_i)(\forall z' \in T^\circ | C)$

$(z <_{T^\circ} z' \rightarrow f(z') < r_i')\}$, $i = 1,\ldots,n$. Set $[\![u]\!] = [\![u]\!]_1 \times \ldots \times [\![u]\!]_n$. Note that $[\![-]\!]$ is definable in $M'[C]$. We clearly have $\tilde{T}_b^p = \{u \in \tilde{T}_b \mid \bigwedge_{i=1}^{n}(b_i \cap [\![u]\!]_i \neq \emptyset)\}$.

Suppose the lemma is false. Then we can find $u_o = \langle x_o, f_o\rangle \in \tilde{T}_b^p$ with no extension in $D \cap \tilde{T}_b^p$. (Let $ht_{\tilde{T}}(u_o) = \theta$.) This is a statement about $M'[b_1,\ldots,b_n][C]$, so by choice of C we can find $R_o \in \mathbb{C}^M$, compatible with C , such that

(1) $R_o \Vdash_{\mathbb{C}^M}^{M'[b_1,\ldots,b_n]} (\forall u \in \mathring{\tilde{T}}_{\check{b}})(u \leq \check{u}_o \;\&\; u \in \mathring{D} \rightarrow \bigvee_{i=1}^{n}(\check{b}_i \cap [\![u]\!]_i = \emptyset)$.

Let $S^\circ = \{\langle\vec{x}\rangle \in (T^\circ)^n \mid \text{for some } r_i' <_Q r_i (\forall\langle\vec{y}\rangle \geq \langle\vec{x}\rangle)[R_o \Vdash_{\mathbb{C}^M}^{M'[b_1,\ldots,b_n]} (ht(\langle\check{\vec{y}}\rangle) \in \mathring{C} \cap \mathring{\gamma}_{\check{\theta}} \rightarrow \bigwedge_{i=1}^{n}(\check{f}_o(\check{y}_i) <_Q \check{r}_i')) \;.\&.\; (\forall u \in \mathring{\tilde{T}}_{\check{b}})(u \leq u_o \;\&\; u \in \mathring{D} \;\&\; ht(\langle\check{\vec{y}}\rangle) = \mathring{\gamma}_{ht(u)}. \rightarrow \langle\check{\vec{y}}\rangle \notin [\![u]\!])]\}$. Thus S° is a subtree of $(T^\circ)^n$. Let S be the set of all $z \in S^\circ$ with successors in S° at all levels of $(T^\circ)^n$. By (1), $(\forall\alpha < \omega_1)(\langle b_1(\gamma_\alpha),\ldots,b_n(\gamma_\alpha)\rangle \in S)$, so S is a non-empty subtree of $(T^\circ)^n$ such that every member of S has uncountably many successors in S . Also, in the definition of S°, the sentence being forced is $\Sigma_o(M')$, so we can replace $\Vdash_{\mathbb{C}^M}^{M'[b_1,\ldots,b_n]}$ by $\Vdash_{\mathbb{C}^M}^{M'}$ here, to conclude that S is an element of M' . Note that, in $M'[C]$ we have (by definition of S) ,

(2) $\langle\vec{x}\rangle \in S \;\&\; u \in u_o \;\&\; u \in D \;\&\; ht(\langle\vec{x}\rangle) = \gamma_{ht(u)} \rightarrow \langle\vec{x}\rangle \notin [\![u]\!]$.

Since $S \in M'$ and T° is Aronszajn in M' , we can apply Corollary VI.8 to obtain a normal function $\tau : \omega_1 \rightarrow \omega_1$ in M' such that $(\forall\nu < \omega_1)(\forall z \in S_{\tau(\nu)})(\beta > \nu \rightarrow S_{\tau(\beta)}^{(z)}$ is well-distributed) . By lemma 4, we may assume that $\tau \in M$. Then, as C is M-generic over $\mathbb{C}^M$, we can apply lemma 1 to find $\alpha_o < \omega_1$ such that $\gamma_{\alpha_o} = \alpha_o$ and $(\forall\beta \geq \alpha_o)(\tau(\gamma_\beta) = \gamma_\beta)$. So, we have:

(3) $\gamma_{\alpha_o} = \alpha_o \;\&\; (\forall\nu \geq \alpha_o)(\forall z \in S_{\gamma_\nu})(\beta > \nu \rightarrow S_{\gamma_\beta}^{(z)}$ is well-distributed).

By Lemma 6, we can find a normal function $\rho : \omega_1 \to \omega_1$ in M such that for all $\nu < \omega_1$, $S|\rho(\nu) = S \cap V_{\rho(\nu)} \in W_{\rho(\nu)}$. Pick $\alpha < \omega_1$ so that $\gamma_{\alpha_1} = \alpha_1$ and $(\forall\beta \geq \alpha_1)(\rho(\gamma_\beta) = \gamma_\beta)$. So, we have:

(4) $\gamma_{\alpha_1} = \alpha_1$ & $(\forall\beta \geq \alpha_1)(S|\gamma_\beta \in W_{\gamma_\beta})$.

Let α be least such that $\gamma_\alpha = \alpha \geq \alpha_0, \alpha_1$ and $\alpha > ht(u_0) = \theta$.

Now, $S_{\alpha+1} \neq \emptyset$, so by the construction of $\tilde{T}_\alpha$, $\tilde{T}_{\alpha+1}$ (and the first clause in the definition of S°) we can find $u = \langle x,f\rangle \leq u_0$ such that $u \in (\tilde{T}_b)_{\alpha+1}$ and, for some $\langle\vec{x}\rangle \in S_{\alpha+1}$, $\langle\vec{x}\rangle \in [\![u]\!]$.

Let $\beta > \alpha$. By (4), $S|\gamma_{\beta+1} \in W_{\gamma_{\beta+1}} \in N_{\gamma_{\beta+1}}$. So, as $N_{\gamma_{\beta+1}}$ is a ZF^--model, $S_{\gamma_\beta} \in N_{\gamma_{\beta+1}}$. Hence, $S_{\gamma_\beta}, S_{\gamma_{\beta+1}} \in \tilde{N}_\beta$ for all $\beta > \alpha$. If follows, by (3) and the generic nature of the construction of $\tilde{T}$, that (by a simple induction on β) for all $\beta > \alpha+1$, if $u' \in (\tilde{T}_b)$, $u' < u$, and $ht(u') = \beta$, then $S_{\gamma_\beta}^{\langle\vec{x}\rangle} \cap [\![u]\!]$ is well-distrubuted.

Let $u' \in \tilde{T}_b$, $u' < u$, $u' \in D$. Set $\beta = ht(u')$. By (2), $\langle\vec{y}\rangle \in S_{\gamma_\beta} \to \langle\vec{y}\rangle \notin [\![u]\!]$, which is to say $S_{\gamma_\beta} \cap [\![u']\!] = \emptyset$. But we have just seen that $S_{\gamma_\beta}^{\langle\vec{x}\rangle} \cap [\![u']\!]$ will be well-distrituted! Hence no extension of u can lie in D, contrary to D being dense in $\tilde{T}_b$. ▌

By examining the proofs of lemmas 4 to 8, we obtain

<u>Lemma 9</u>

The conclusions of lemmas 4 to 8 are valid whenever M is a countable transitive model of ZF^- + "there is an uncountable cardinal" + $\Diamond^*$. ▌

At last we can prove:

Lemma 10

$M[C] \models$ "$\tilde{T}$ is a Souslin tree" .

Proof: Let J be the usual Σ_1 function such that for any set $\mathfrak{A}$, $\langle J(\alpha,\mathfrak{A}) \mid \alpha \in On \rangle$ enumerates $L[\mathfrak{A}]$.

Let X be a maximal antichain of $\tilde{T}$ in $M[C]$. Since $\tilde{T}, X \in M[C]$, there is a set $Y \subseteq \omega_1$, $Y \in M$, such that $\tilde{T}, X \in L[Y,C]$. We can assume that Y is chosen so that $\omega_1 = \omega_1^{L[Y]}$ and $T, \mathring{T}, W \in L[Y]$. Let τ_0 be the least ordinal such that $\langle X,\tilde{T}\rangle = J(\tau_0,\langle Y,C\rangle)$, and set τ_1 be least such that $\langle T,\mathring{T},W\rangle = J(\tau_1,Y)$.

Applying lemma 3 we obtain $N \in M$, $N \prec H^M_{\omega_2}$, $|N|^M = \omega$, such that $\langle Y,\tau_0,\tau_1\rangle \in N$, and a countable transitive set $U \in M$, such that, setting $\alpha = \alpha_N$ and $\pi = \pi_N$,

(i) $\bar{N}$, $T|(\alpha+1)$, $\langle \mathring{T}_y \mid y \in T_{\eta(\alpha+1)}\rangle \in U$;

(ii) $C \cap \alpha$ is U-generic over $\pi(\mathbb{C}^M)$;

(iii) there is a map $\pi' \in M[C]$, $\pi' \supseteq \pi$, such that $\pi'(C) = C \cap \alpha$ and $\pi'^{-1} : \bar{N}[C\cap\alpha] \prec H^{M[C]}_{\omega_2}$.

Clearly, $\pi(T) = T|\alpha$, $\pi(W) = W\restriction\alpha$, and $\gamma_\alpha = \alpha$. Since $\tau_0, Y, C \in \mathrm{dom}(\pi')$ and $\mathrm{dom}(\pi') \prec H^{M[C]}_{\omega_2}$, $X,\tilde{T} \in \mathrm{dom}(\pi')$. Clearly, $\pi'(\tilde{T}) = \tilde{T}|\alpha$ and $\pi'(X) = X_\alpha =_{Def} X \cap (\tilde{T}|\alpha)$. Also, X_α is clearly a maximal antichain of $\tilde{T}|\alpha$. Set $D = \{y \in \tilde{T}|\alpha \mid (\exists x \in X_\alpha)(y \leq_{\tilde{T}} x)\}$. Then $D \in \bar{N}[C\cap\alpha]$ is a dense subset of $\tilde{T}|\alpha$.

For each $x \in T_\alpha$, set $b_x = \{y \in T|\alpha \mid x <_T y\}$, a cofinal branch of $T|\alpha$. Now, $T|\alpha$ is a Souslin tree in $\bar{N}$, so by lemma 9 we can apply lemma 7 to $\bar{N}[C\cap\alpha]$, $\tilde{T}|\alpha$, b_x, D, $\tilde{T}_{b_x} =_{Def} \{\langle x',f'\rangle \in \tilde{T}|\alpha \mid x' \in b_x\}$, to conclude that the set $D \cap \tilde{T}_{b_x}$ is dense in $\tilde{T}_{b_x}$, each $x \in T_\alpha$.

Again, for each $x \in T_\alpha$, let B_x be the set of all branches b of $\mathring{T}_x$ of the form $b = \{z \in \mathring{T}_y \mid w <_{\mathring{T}} z\}$ for some $y \in T_{\eta(\alpha+1)}$, $y <_T x$, $w \in (\mathring{T}_y)_\alpha$. Let $b_1,\dots,b_n \in B_x$, and let $p : \{b_1,\dots,b_n\} \to \mathbb{Q}$ be arbitrary. By choice of U, $C \cap \alpha$ is $\bar{N}[b;b_1,\dots,b_n]$-generic over $\mathbb{C}^{\bar{N}}$, so by lemma 9 we can apply lemma 8 to $\bar{N}, b_x, T|\alpha$, $\mathring{T}_x$ (and $\pi(\mathring{T})$), $b_1,\dots,b_n$, p, $C \cap \alpha$, $\mathbb{C}^N$, $\tilde{T}|\alpha$, $D \cap \tilde{T}_{b_x}$, to conclude that $D \cap \tilde{T}^p_{b_x}$ is dense in $\tilde{T}^p_{b_x}$, each $x \in T_\alpha$, p as above.

Claim. $D \in \tilde{N}_\alpha$. (We prove this later.)

Now, $\gamma_\alpha = \alpha$ (and $\lim(\alpha)$), so $\tilde{T}_\alpha$ was constructed generically over $\tilde{N}_\alpha$ for the posets $\mathbb{P}^x$, $x \in T_\alpha$. For such x, let $D^x = \{\langle u,p\rangle \in \mathbb{P}^x \mid u \in D\}$. By the claim, $D^x \in \tilde{N}_\alpha$. By our arguments above, D^x is dense in $\mathbb{P}^x$. Thus, if $u \in \tilde{T}_\alpha$, $u \leq_{\tilde{T}} u'$ for some $u' \in D$, whence u extends an element of X_α. Hence X_α is a maximal antichain in $\tilde{T}$, proving that $X = X_\alpha$, which is countable. There remains only the proof of the claim.

Let $\rho = \omega_2^{L[Y,C]}$. Set $\rho' = \pi''(\rho \cap N)$, $N' = L_{\rho'}[Y \cap \alpha, C \cap \alpha]$. Then, either $\rho < \omega_2$ and $\rho' = \pi(\rho)$, or else $\rho = \omega_2$ and $\rho' = On \cap \bar{N}$. Whichever of these is the case, we must have $N' \subseteq \bar{N}$ and $(\pi'^{-1} \restriction N') : N' \prec L_\rho[Y,C]$. Since $L_\rho[Y,C]$ is closed under J and $\tau_0 < \rho$, $X, \tilde{T} \in L_\rho[Y,C]$. Hence X_α, $\tilde{T}|\alpha \in N'$. In particular, therefore, $D \in N'$, and it suffices to prove that $N' \subseteq \tilde{N}_\alpha$.

Now, $Y, W \in N$, so there is a set $A \in N$ such that $N \models$ "A is closed and unbounded in ω_1 & $\nu \in A \to Y \cap \nu \in W_\nu$". Since $\pi(A) = A \cap \alpha$, $A \cap \alpha$ is closed and unbounded in α, whence $\alpha \in A$. Thus, as $N \prec H^M_{\omega_2}$, we see that $Y \cap \alpha \in W_\alpha$. And, recall that $\gamma_\alpha = \alpha$ (because $\pi'(\langle \gamma_\nu \mid \nu < \omega_1\rangle) = \langle \gamma_\nu \mid \nu < \alpha\rangle$ is cofinal in α.) So, as $|W_\alpha|^{L[Y]} \leq \omega$, $N' = L_{\rho'}[Y \cap \alpha, C \cap \alpha] \subseteq L_{\rho''}[W_\alpha, C \cap \alpha]$, where $\rho'' = \max(\rho', \delta(\alpha))$.

Thus, if $\rho' < \delta(\alpha)$, then $N' \subseteq L_{\delta(\alpha)}[W_\alpha, C \cap \alpha] \subseteq N_\alpha \subseteq \tilde{N}_\alpha$, as required. We prove that $\rho' < \delta(\alpha)$.

Suppose first that α is countable in $L[Y \cap \alpha, C \cap \alpha]$. Then α is countable in $L_{\delta(\alpha)}[Y \cap \alpha, C \cap \alpha]$, by definition of δ . But $\alpha = \omega_1^{N'}$. Hence $\rho' < \delta(\alpha)$.

Now suppose that $\alpha \geq \omega_1^{L[Y\cap\alpha, C\cap\alpha]}$ but $(\alpha^+)^{L[Y\cap\alpha, C\cap\alpha]} < \omega_1$. Again by definition of δ , we have $\rho' \leq \omega_2^{L[Y\cap\alpha, C\cap\alpha]} < \delta(\alpha)$ (the first inequality here is easily established, since $\rho' = \pi''(\rho \cap N)$.)

Finally, we show that there are no further possibilities. Well, if $(\forall\beta < \omega_1)(\omega_1^{L[Y\cap\beta, C\cap\beta]} < \omega_1)$, then $(\forall\beta < \omega_1)(\omega_1$ is inaccessible in $L[Y \cap \beta, C \cap \beta])$, so the above cases suffice. Otherwise, let β be least such that $\omega_1^{L[Y\cap\beta, C\cap\beta]} = \omega_1$. Since β is thus $H_{\omega_2}^{M[C]}$-definable, we have $\beta \in \mathrm{dom}(\pi')$, whence $\beta < \alpha$. Then, if α is uncountable in $L[Y \cap \alpha, C \cap \alpha]$, we have $\alpha \geq \omega_1^{L[Y\cap\beta, C\cap\beta]} = \omega_1$, which is absurd. ▌

Chapter X

$\mathrm{CON}(\mathrm{ZF}) \rightarrow \mathrm{CON}(\mathrm{ZFC} + \mathrm{GCH} + \mathrm{SH})$

We piece together all of our previous results to prove the following theorem:

Theorem 0 (Jensen)

Let M be a c.t.m. of $\mathrm{ZFC} + (2^{\omega} = \omega_1) + (2^{\omega_1} = \omega_2)$. There is a cardinal absolute generic extension N of M such that:

(i) For all cardinals λ of M, $(2^{\lambda})^N = (2^{\lambda})^M$;

(ii) $\mathcal{P}^N(\omega) = \mathcal{P}^M(\omega)$;

(iii) $N \models \mathrm{SH}$.

For the whole of the chapter, then, M will be as above. By the results of chapter VIII we shall also assume that $M \models \Box \;\&\; \Diamond^*$, so the iteration lemma holds in M. We first require a lemma concerning the closed set forcing poset, $\mathbb{C}$, of the previous chapter.

Lemma 1

There is a Cohen extension M^* of M such that:

(i) M and M^* have the same cardinals and cofinality function;

(ii) for all cardinals $\varkappa$ of M, $(2^{\varkappa})^{M^*} = (2^{\varkappa})^M$;

(iii) $\mathcal{P}^{M^*}(\omega) = \mathcal{P}^M(\omega)$;

(iv) $M^* = M[\langle C_\nu \mid \nu < \omega_2^M \rangle]$, where, for each $\nu < \omega_2^M$, if we set $M_\nu = M[\langle C_\tau \mid \tau < \nu \rangle]$, C_ν is an M_ν-generic subset of $\omega_1^{M_\nu}$ over $\mathbb{C}^{M_\nu}$;

(v) $M^* \models \Box + \Diamond$;

(vi) for all $\alpha < \omega_2^M$, $M_\alpha \models \Box + \Diamond^*$;

(vii) if $\alpha < \omega_2^M$ and T is a Souslin tree in M_α, then T is a Souslin tree in M^*.

Proof: This is a classic example of the kind of forcing described in chapter VII. We iterate the closed set algebra ω_2 times in M. At limit stages of cofinality ω we take (what will at first appear to be) a simple modification of the inverse limit (we describe this "modification" below); at limit stages of cofinality ω_1 we take the direct limit. Let $\mathbb{B}$ be the direct limit of the entire iteration sequence so constructed. It will be clear that $\mathbb{B}$ will have a dense subset which is ω_1-closed, so if M^* is a generic extension of M determined by $\mathbb{B}$, then (iii) and (iv) are immediate. We must check that (i) holds, whence (ii) will be immediate of course. It suffices to show that $\mathbb{B}$ satisfies ω_2-c.c. (in M). Well, by a straightforward generalisation (from ω_1 to ω_2) of the argument of Solovay-Tennenbaum described in [Je 1], Theorem 49, we see that if each stage in the iteration has ω_2-c.c., then so does $\mathbb{B}$. And clearly, the ω_2-c.c. will be preserved at limit stages of cofinality ω_1. Successor stages also cause no problem, since $\mathbb{C}$ has ω_2-c.c. [For suppose we are given ω_2 members of $\mathbb{C}$. We can assume, by discarding some of them if necessary, that they are of the form $\langle \nu, A_\alpha \rangle$, $\alpha < \omega_2$, for some fixed $\nu < \omega_1$. We may likewise assume that $\alpha < \beta < \omega_2 \rightarrow A_\alpha \cap \nu = A_\beta \cap \nu$. But then we see that for any $\alpha < \beta < \omega_2$, $\langle \nu, A_\alpha \cap A_\beta \rangle$ extends both $\langle \nu, A_\alpha \rangle$ and $\langle \nu, A_\beta \rangle$, so we are through.] It remains to check that ω_2-c.c. is preserved at limit stages of cofinality ω. It is in order to facilitate this argument that we (ostensibly) modify the definitions of the inverse limit as it was described in chapter VII.

Suppose then α is a limit ordinal, $\mathrm{cf}(\alpha) = \omega$, and that we have defined $\langle \mathbb{P}_\nu | \nu < \alpha \rangle$, $\langle \mathbb{B}_\nu | \nu < \alpha \rangle$, $\langle e_{\nu\tau} | \nu < \tau < \alpha \rangle$, $\langle h_{\tau\nu} | \nu < \tau < \alpha \rangle$

(where we use the same notation as in Chapter VII ; thus $\mathbb{P}_{\nu+1}$ corresponds to $\mathbb{P}_\nu$, $\mathbb{B}_\nu$, $\mathbb{C}^{M^{\mathbb{B}_\nu}}$ via lemma VII.2, etc.). For $\mathbb{P}_\alpha$, we take the set of all sequences $p = \langle p_\nu | \nu < \alpha \rangle$ such that: (i) $\nu < \alpha \to p_\nu \in \mathbb{P}_\nu$; (ii) $\nu < \tau < \alpha \to p_\nu = h_{\tau\nu}(p_\tau)$; (iii) for some countable set $\chi \subseteq \alpha$, if $\tau < \alpha$ is such that $\nu < \tau \to e_{\nu\tau}(p_\nu) \neq p_\tau$, then either $p_\tau = \bigwedge_{\nu<\tau} e_{\nu\tau}(p_\nu)$ (in $\mathbb{B}_\tau$) or else $\tau \in \chi$. (We call χ a <u>support</u> for p in such a situation.) Since we have (it appears) modified the idea of an inverse limit, we should check that ω_1-happiness is preserved at stage α . Clearly, the only possible source of trouble is condition (iii) of the neat cover requirements. But to verify this, all we need do is carry out a simple inductive construction of length α ; and by the very definition of the iteration sequence, limit stages in such a construction cause no problems.

We shall prove that the ω_2 - c.c. is preserved under this limit operation. Before we do so however, let us now take the time off to <u>mention</u> that it would make no difference whatsoever to the construction if we defined $\mathbb{P}_\alpha$ as the inverse limit of a cofinal ω-subsequence of $\langle \mathbb{P}_\nu | \nu < \alpha \rangle$, as in chapter VII. More precisely, if one defines $\langle \mathbb{P}_\nu | \nu < \omega_2 \rangle$ as above and $\langle \mathbb{P}'_\nu | \nu < \omega_2 \rangle$ using the "original" ω-inverse limit instead, then by an easy induction on $\nu < \omega_2$, $\mathbb{P}_\nu \cong \mathbb{P}'_\nu$. It is, however, sufficient and slightly more convenient, to stick to the approach we have adopted, and the above equivalence will therefore not be required.

Assume then that $\mathbb{P}_\nu$ satisfies ω_2 - c.c. for each $\nu < \alpha$. We show that $\mathbb{P}_\alpha$ satisfies ω_2 - c.c. For clarity, we consider the case $\alpha = \omega$ first, and then indicate how the argument may be modified for the more general case.

In ZFC , we may define functions f, h on $\mathbb{C}$ thus: if $p = \langle \nu, A\rangle$, let $f(p) = \nu$, $h(p) = \nu \cap A$.

For $p \in {}^{\omega}V$, define $\sigma_i(p)$, $i < \omega$, by induction, thus:

$\sigma_0(p) = p_0$; $\sigma_1(p) = \langle p_0, p_1\rangle$; $\sigma_{i+1}(p) = \langle \sigma_i(p), p_{i+1}\rangle$.

If $p \in {}^{n+1}V$, define $\sigma_0(p), \ldots, \sigma_n(p)$ similarly.

Recall now the <u>proof</u> of lemma VII.2.

For each $n < \omega$, let $\mathbb{P}^{\#}_n = \{p = \langle p_0, \ldots, p_n\rangle \mid (\forall i \leq n)(\sigma_i(p) \in \mathbb{P}_i)\}$. Set $p \leq^{\#}_n q \leftrightarrow \sigma_n(p) \leq_n \sigma_n(q)$.

Thus $\sigma_n : \mathbb{P}^{\#}_n \cong \mathbb{P}_n$, and clearly $\mathbb{P}^{\#}_n$ is a sort of representation of $\mathbb{P}_n$ in a form as if lemma VII.2 had not been involved at all (ie. to obtain $\mathbb{P}^{\#}_n$ from $\mathbb{P}_n$ we unravel the definition (using VII.2) of $\mathbb{P}_n, \mathbb{P}_{n-1}, \ldots, \mathbb{P}_1$).

For each $n < \omega$, let $\mathbb{P}^{*}_n = \{p \in \mathbb{P}^{\#}_n \mid (\exists \nu < \omega_1)(\exists X_0, \ldots, X_n \subseteq \nu)$ $[f(p_0) = \nu \ \& \ h(p_0) = X_0 \ \& \ \bigwedge_{0 \leq i < n} [\sigma_i(p) \Vdash_i \text{"} f(p_{i+1}) = \check{\nu} \ \& \ h(p_{i+1}) = \check{X}_{i+1}\text{"}]]\}$.

<u>Claim 1</u>. $\mathbb{P}^{*}_n$ is neatly ω_1-closed in $\mathbb{P}^{\#}_n$.

This is proved by induction on n . For $n = 0$ it is trivial. Assume it holds for n . Let $\langle p^m \mid m < \omega\rangle$ be a decreasing sequence from $\mathbb{P}^{*}_{n+1}$. For each $m < \omega$, let ν_m, X_m be such that $\sigma_n(p^m) \Vdash_n \text{"} f(p^m_{n+1}) = \check{\nu}_m \ \& \ h(p^m_{n+1}) = \check{X}_m\text{"}$. Set $\nu = \sup_{m<\omega} \nu_m$, $X = \bigcap_{m<\omega} X_m$. Now, by induction hypothesis, if $q = \bigwedge_{m<\omega} \sigma_n(p^m)$ (in $BA(\mathbb{P}^{\#}_n)$) , then $q \in \mathbb{P}^{\#}_n$. Since we clearly have $q \Vdash_n \text{"}\langle p^m_{n+1} \mid m < \omega\rangle$ is a decreasing sequence from $\mathbb{C}$", use the maximum principle to find a unique r such that $q \Vdash_n$ "$r = \bigwedge_{m<\omega} p^m_{n+1}$ (in $\mathbb{C}$)". Clearly, $q \Vdash_n \text{"} f(r) = \check{\nu} \ \& \ h(r) = \check{X}\text{"}$. Hence $p = \langle q, r\rangle \in \mathbb{P}^{*}_{n+1}$. But clearly, $p = \bigwedge_{m<\omega} p^m$ (in $BA(\mathbb{P}^{\#}_{n+1})$), so we are done.

<u>Claim 2</u>. $\mathbb{P}_n^*$ is dense in $\mathbb{P}_n^\#$.

We prove claim 2 by induction on n. For $n = 0$ there is nothing to prove. Assume it holds for n. Let $p \in \mathbb{P}_{n+1}^\#$. By induction hypothesis and claim 1, we can find $p^o \in \mathbb{P}_n^*$, $p^o \leq_n^\# p \restriction n+1$, such that $\sigma_n(p^o) \Vdash_n$ "$f(p_{n+1}) = \check{\nu}_o$ & $h(p_{n+1}) = \check{X}_o$" for some $\nu_o < \omega_1$, $X_o \subseteq \nu_o$. Let $\nu_1 = f(p_o^o)$. We may clearly assume $\nu_1 \geq \nu_o$ here. Again, we can find $p^1 \in \mathbb{P}_n^*$, $p^1 \leq_n^\# p^o$, and q^1 such that $\sigma_n(p^1) \Vdash_n$ "$q^1 \leq p_{n+1}$ (in $\mathbb{C}$) & $f(q^1) = \check{\nu}_1$ & $h(q^1) = \check{X}_1$" for some $X_1 \subseteq \nu_1$. Let $\nu_2 = f(p_o^1)$. We may assume $\nu_2 \geq \nu_1$. Similarly, pick $p^2 \in \mathbb{P}_n^*$, $p^2 \leq_n^\# p^1$, and q^2 such that $\sigma_n(p^2) \Vdash_n$ "$q^2 \leq q^1$ (in $\mathbb{C}$) & $f(q^2) = \check{\nu}_2$ & $h(q^2) = \check{X}_2$" for some $X_2 \subseteq \nu_2$. Proceed inductively now, and finish by setting $p^\omega = \bigwedge_{i<\omega} p^i$ (in $BA(\mathbb{P}_n^\#)$). By claim 1, $p^\omega \in \mathbb{P}_n^*$. Since $\sigma_n(p^\omega) \Vdash_n$ "$\langle q^i | i < \check{\omega}\rangle$ is a decreasing seqience from $\mathbb{C}$", we can find a unique q^ω such that $\sigma_n(p^\omega) \Vdash_n$ "$q^\omega = \bigwedge_{i<\omega} q^i$ (in $\mathbb{C}$)". Setting $\nu = \sup_{i<\omega} \nu_i$, $X = \bigcap_{i<\omega} X_i$, we clearly have $\sigma_n(p^\omega) \Vdash_n$ "$f(q^\omega) = \check{\nu}$ & $h(q^\omega) = \check{X}$". But clearly, $f(p_o^\omega) = \nu$. Hence $\langle p^\omega, q^\omega\rangle \in \mathbb{P}_{n+1}^*$. Since $\langle p^\omega, q^\omega\rangle \leq_{n+1}^\# p$, we are done.

Set $\mathbb{P}_\omega^\# = \{p = \langle p_n | n < \omega\rangle \mid (\forall n < \omega)(p \restriction n+1 \in \mathbb{P}_n^\#\}$, and let $p \leq_\omega^\# q \leftrightarrow (\forall n < \omega)(p \restriction n+1 \leq_n^\# q \restriction n+1\}$. Define $\sigma : \mathbb{P}_\omega^\# \to \mathbb{P}_\omega$ by $\sigma(p) = \langle \sigma_n(p) | n < \omega\rangle$. Clearly, $\sigma : \mathbb{P}_\omega^\# \cong \mathbb{P}_\omega$. So, to prove that $\mathbb{P}_\omega$ has ω_2-c.c., it suffices to show that $\mathbb{P}_\omega^\#$ does. Set $\mathbb{P}_\omega^* = \{p \in \mathbb{P}_\omega^\# \mid (\forall n < \omega)(p \restriction n+1 \in \mathbb{P}_n^*)\}$.

<u>Claim 3</u>. $\mathbb{P}_\omega^*$ is dense in $\mathbb{P}_\omega^\#$.

Let $p \in \mathbb{P}_\omega^\#$ be given. Set $p^o = p$. Using claim 2, pick $p^1 \leq_\omega^\# p^o$ such that $p^1 \restriction 2 \in \mathbb{P}_1^*$; and in general pick $p^{n+1} \leq_\omega^\#$

p^n such that $p^{n+1}\lceil n+2 \in \mathbb{P}^*_{n+1}$. For each $n < \omega$, let $q_n = \bigwedge_{i<\omega} p^i \lceil n+1$ (in $BA(\mathbb{P}^\#_n)$) . By claim 1, $q_n \in \mathbb{P}^*_n$, each n . Define $q \in {}^\omega V$ by $q\lceil n+1 = q_n$, each $n < \omega$. Clearly, $q \in \mathbb{P}^*_\omega$ and $q \leq^\#_\omega p$, as required.

Now let A be a pairwise incompatible subset of $\mathbb{P}^\#_\omega$ of cardinality ω_2 . By claim 3 we may assume that $A \subseteq \mathbb{P}^*_\omega$. We may further assume that $f(p_o) = \nu$ for all $p \in A$, where ν is fixed here. For each $p \in A$, let $\langle X_i(p) | i < \omega \rangle$ be such that $h(p_o) = X_o(p)$ and $(\forall i < \omega)(\sigma_i(p) \Vdash_i$ "$h(p_{i+1}) = \widetilde{X_{i+1}(p)}$".) As p vaires through A , $\langle X_i(p) | i < \omega \rangle$ varies through ${}^\omega \mathcal{P}(\nu)$. Hence, as $|{}^\omega \mathcal{P}(\nu)| = 2^{\omega \cdot \omega} = \omega_1$, we may assume that for all $p \in A$, $(\forall i < \omega)(X_i(p) = X_i)$, where $\langle X_i | i < \omega \rangle$ is fixed. But clearly, under these conditions, A is pairwise compatible, which is absurd. The proof is complete.

It is easily seen how to modify the above proof for the case $\lim(\alpha)$, $\omega < \alpha < \omega_1$. [Perhaps the easiest way to express a formal proof is to proceed by induction on the limit ordinals $\alpha < \omega_1$, establishing analogues of claims 1, 2, and 3 at each step.]

Suppose now that $\lim(\alpha)$, $cf(\alpha) = \omega$, and $\omega_1 < \alpha < \omega_2$. Let A be a pairwise incompatible subset of $\mathbb{P}_\alpha$ of cardinality ω_2. For each $p \in A$, let $s(p)$ be a countable support for p . As p varies through A , there are at most $\omega_1^\omega = \omega_1$ possibilities for $s(p)$. Hence we may assume $s(p) = s$ for all $p \in A$, where s is a fixed countable (and cofinal, by the induction hypothesis that $\mathbb{P}_\nu$ satisfies ω_2-c.c. for all $\nu < \alpha$) subset of α . But since we need only now concern ourselves with the sub-poset $\{p \in \mathbb{P}_\alpha |$ support $(p) = s\}$, and since $otp(s)$ is a countable limit ordinal, we can obtain the desired contradiction by an argument as above (eg. by induction "along" s).

That completes the proof of lemma 1 (i). We must check that (v)-(vii) are valid. Well (v) holds by lemmas VIII.2 and VIII.6 , (vi) by lemmas VIII.2 and VIII.4 , and (vii) by lemma VII.4. The lemma is proved. ∎

Using the notation of lemma 1, we can restate the "crucial lemma" of the previous chapter as follows:

Lemma 2

Let $\alpha < \omega_2^M$ and let $\mathbb{B}$ be a Souslin algebra in M_α . Let $\mathring{T} \in M_\alpha^{(\mathbb{B})}$ be, with $\mathbb{B}$-value $\mathbb{1}$, an Aronszajn tree. Then there is a Souslin algebra $\mathbb{B}'$ in $M_{\alpha+1}$ such that $\mathbb{B}$ is a nice subalgebra of $\mathbb{B}'$ and $\|\mathring{T}$ is special$\|^{\mathbb{B}} = \mathbb{1}$. ∎

We shall in fact carry out our main iteration within M^* . (By lemma 1, this will be quite in order). The next lemma adapts lemma 2 to this situation.

Lemma 3

Assume $V = M^*$. Let $\mathbb{B}$ be a Souslin algebra and let $\mathring{T} \in V^{(\mathbb{B})}$ be, with $\mathbb{B}$-value $\mathbb{1}$, an Aronszajn tree. Then there is a Souslin algebra $\mathbb{B}'$ such that $\mathbb{B}$ is a nice subalgebra of $\mathbb{B}'$ and $\|\mathring{T}$ is special$\|^{\mathbb{B}'} = \mathbb{1}$.

Proof: Let U be an arbitrary Souslinisation of $\mathbb{B}$. We may assume $U, \mathbb{B} \subseteq H_{\omega_1}$. Thus for some $\alpha < \omega_2$, $U, \mathbb{B}, \mathring{T} \in M_\alpha$. Then, since U is a Souslin tree in M_α , we see that $\mathbb{B}$ is a Souslin algebra in M_α and $\mathring{T} \in M_\alpha^{(\mathbb{B})}$. By lemma 2 we can find a Souslin algebra $\mathbb{B}'$ in $M_{\alpha+1}$ such that $\mathbb{B}$ is a nice subalgebra of $\mathbb{B}'$ and $\|\mathring{T}$ is special$\|^{\mathbb{B}'} = \mathbb{1}$. Let U' be an arbitrary Souslinisation of $\mathbb{B}'$ in $M_{\alpha+1}$. By lemma 1 (vii), U' will really be a Souslin tree (ie. in M^*), so $\mathbb{B}'$ will be a Souslin algebra (ie. in M^*) and $\mathbb{B}$ will still be (in M^*) a nice subalgebra of $\mathbb{B}'$. Since $\|\mathring{T} \text{ is special}\|^{\mathbb{B}'} = \mathbb{1}$ in $M_{\alpha+1}$, we clearly have $\|\mathring{T} \text{ is special}\|^{\mathbb{B}'} = \mathbb{1}$, so we are done.▌

The next lemma proves Theorem 0.

<u>Lemma 4</u>

There is a BA $\mathbb{B}^*$ in M^* such that

(i) $M^* \models$ "$|\mathbb{B}^*| = \omega_2$ & $\mathbb{B}^*$ satisfies c.c.c. & $\mathbb{B}^*$ is ω_1-dense";

(ii) if G is M^*-generic for $\mathbb{B}^*$, then $M^*[G] \models SH$.

Proof: Assume $V = M^*$. Define a function f so that for any Souslin algebra $\mathbb{B}$, the sequence $\langle f(\mathbb{B},\nu) \mid \nu < \omega_2\rangle$ enumerates $\{\mathring{X} \in V^{(\mathbb{B})} \mid \|\mathring{X} \subseteq \check{\omega}_1 \times \check{\omega}_1\|^{\mathbb{B}} = \mathbb{1}\}$. Let $\prec -,-\succ$ be a bijective pairing function on ω_2 such that for all $\alpha,\beta < \omega_2$, $\alpha,\beta \leq \prec\alpha,\beta\succ < \omega_2$. Let $(-)_0$, $(-)_1$ be the inverse functions to $\prec -,-\succ$. Define a function σ as follows. Since $\Diamond$ holds we can let $\sigma(\mathbb{2},\emptyset)$ be an arbitrary Souslin algebra. Suppose now that $\nu < \omega_2$ and $\langle \mathbb{B}_\tau \mid \tau \leq \nu\rangle$ is a sequence of Souslin algebras with $\mathbb{B}_\tau$ a nice subalgebra of $\mathbb{B}_\eta$ for each $\tau < \eta \leq \nu$. If $f(\mathbb{B}_{(\nu)_0}, (\nu)_1)$ is, with $\mathbb{B}_\nu$ value $\mathbb{1}$, an Aronszajn tree in $V^{(\mathbb{B}_\nu)}$, then we obtain $\sigma(\mathbb{B}_\nu, \langle \mathbb{B}_\tau \mid \tau < \nu\rangle)$ from $\mathbb{B}_\nu$ and $f(\mathbb{B}_{(\nu)_0}, (\nu)_1)$ by lemma 3. Otherwise we let $\sigma(\mathbb{B}_\nu, \langle \mathbb{B}_\tau \mid \tau < \nu\rangle) = \mathbb{B}_\nu$.

This defines a function σ for which we can apply the main iteration lemma, lemma VIII.12. Let $\langle \mathbb{B}_\nu \mid \nu < \omega_2\rangle$ be the sequence which lemma VIII.12 gives us in this case. Let $\mathbb{B}^* = \bigcup_{\nu<\omega_2} \mathbb{B}_\nu$. Then (i) is immediate. We prove (ii). Suppose $\|\text{There is a non-special Aronszajn tree}\|^{\mathbb{B}^*} > 0$. Now

$\|(\exists X)[X \subseteq \check{\omega}_1 \times \check{\omega}_1$ and if there is a non-special Aronszajn tree then X is one$)\|^{\mathbb{B}^*} = \mathbb{1}$, so by the maximum principle for $V^{(\mathbb{B}^*)}$ we can pick $\mathring{X} \in V^{(\mathbb{B}^*)}$ such that $\|\mathring{X} \subseteq \check{\omega}_1 \times \check{\omega}_1\|^{\mathbb{B}^*} = \mathbb{1}$ and (in particular) $\|\mathring{X}$ is a non-special Aronszajn tree$\|^{\mathbb{B}^*} > \mathbb{0}$. Pick $\eta < \omega_2$ so that $\mathring{X} \in V^{(\mathbb{B}_\eta)}$. Then $\|\mathring{X} \subseteq \check{\omega}_1 \times \check{\omega}_1\|^{\mathbb{B}_\eta} = \mathbb{1}$ so $\mathring{X} = f(\mathbb{B}_\eta, \tau)$ for some $\tau < \omega_2$, Let $\nu = \langle \eta, \tau \rangle$. Then, clearly $\|\mathring{X}$ is an Aronszajn tree$\|^{\mathbb{B}_\nu} = \mathbb{1}$, so by construction, $\|\mathring{X}$ is special$\|^{\mathbb{B}_{\nu+1}} = \mathbb{1}$. Hence, $\|\mathring{X}$ is special$\|^{\mathbb{B}^*} = \mathbb{1}$, of course, which is a contradiction. Thus $\|$Every Aronszajn tree is special$\|^{\mathbb{B}^*} = \mathbb{1}$, and we are done. ∎

Appendix

ITERATED SOUSLIN FORCING AND $\mathcal{P}(\omega)$

In [Jn Jo], Jensen and Johnsbråten give a generic construction of a non-constructible Δ^1_3 set of integers by means of an ω-iteration of Souslin forcing. Their construction is a specialisation of an earlier construction which Jensen developed in order to show that iterated Souslin forcing over L must introduce new reals. For the convenience of the reader, and because the construction in [Jn Jo] is (necessarily) fairly cumbersome, we outline here a simplification which just gives the above result.

In L, we define a sequence $\langle T^n \mid n<\omega\rangle$ of ω_1-trees such that:

(i) T^0 is Souslin;

(ii) if b_0 is an L-generic branch of T^0, then we can canonically define a subtree $\hat{T}^1$ of T^1 in $L[b_0]$ such that $\hat{T}^1$ is Souslin in $L[b_0]$;

(iii) if b_1 is an $L[b_0]$-generic branch of $\hat{T}^1$, then we can canonically define a subtree $\hat{T}^2$ of T^2 in $L[b_0,b_1]$ such that $\hat{T}^2$ is Souslin in $L[b_0,b_1]$;

(iv)$\cdots<(\omega)$ similar;

(ω) if $\langle b_0,b_1,\dots\rangle$ is any such sequence of branches of the respective trees $T^0,\hat{T}^1,\dots$, then $L[\langle b_0,b_1,\dots\rangle] = L[a]$ for some $a \subseteq \omega$.

This shows that, regardless of what we do at limit stages, if we try to obtain SH by forcing over L, iteratively destroying Souslin trees, if the method we use to destroy Souslin trees <u>results</u> in their being given a branch, then the iteration must introduce new reals.

As we proceed, given $x \in T^n$ we shall define a subtree t_x of T^{n+1}, such that:

(i) t_x is a $ht(x)$-tree;

(ii) $x \leq_n y \rightarrow t_x = t_y | ht(x)$;

(iii) if x, y are incomparable in T^n and z is the largest $z \leq_n x, y$ in T^n, then $t_x \cap t_y = t_z$ (unless $ht(z) = 0$, when $t_x \cap t_y = \{\langle n+1 \rangle\}$);

(iv) $T^{n+1} = \bigcup_{x \in T^n} t_x$.

Each T^n_α will consist of α-sequences of natural numbers, and the ordering, $\leq_n$, will be the usual one. We define $T^0|1$ first, then $T^1|1$, then $T^2|1, \ldots$, then $T^0|2$, then $T^1|2$, then $T^2|2, \ldots$, etc.

Fix some canonical, recursive bijection $\{-,-\} : \omega \times \omega \leftrightarrow \omega$.

Stage 0. For each $n \in \omega$, set $T^n_0 = \{\langle n \rangle\}$, $t_{\langle n \rangle} = \emptyset$.

Stage 1. For each $n \in \omega$, set $T^n_1 = \{\langle n,i \rangle \mid i \in \omega\}$, $t_{\langle n,i \rangle} = \{\langle n+1 \rangle\}$.

Stage $\alpha+2$. For each $n \in \omega$, set $T^n_{\alpha+2} = \{x^\frown\langle i \rangle \mid x \in T^n_{\alpha+1} \ \& \ i \in \omega\}$, and for $x \in T^n_{\alpha+1}$, $i \in \omega$, set $t_{x^\frown\langle i \rangle} = t_x \cup \{y^\frown\langle \{i,j\} \rangle \mid y \in t_x \ \& \ ht(y) = \alpha \ \& \ j \in \omega\}$.

Stage α, lim(α). We first define T^0_α. Let $\eta = \eta_{0,\alpha}$ be least such that $T^0|\alpha \in L_\eta$, $L_\eta \models ZF^-$, and $|\alpha|^{L_\eta} = \omega$. For each $x \in T^0|\alpha$, let b_x be the $<_L$-least L_η-generic branch of $T^0|\alpha$ through x, and let $T^0_\alpha = \{\bigcup b_x \mid x \in T^0|\alpha\}$. Assume now that T^n_α is defined. Define t_x for $x \in T^n_\alpha$ by $t_x = \bigcup_{y <_n x} t_y$. Let $\eta = \eta_{n+1,\alpha}$ be least such that $T^n|\alpha+1$, $T^{n+1}|\alpha$, $\langle t_x \mid x \in T^n_\alpha \rangle \in L_\eta$, $L_\eta \models ZF^-$, and $|\alpha|^{L_\eta} = \omega$. For each $x \in T^n_\alpha$, define a poset $\mathbb{P} = \mathbb{P}_x$ in L_η as follows. The elements of $\mathbb{P}$ are finite maps f with $dom(f) \subseteq \omega$ and $ran(f) \subseteq t_x$, such that $i,j \in dom(f) \rightarrow ht(f(i)) = ht(f(j))$. The ordering is defined by $f' \leq f \leftrightarrow dom(f') \supseteq dom(f) \ \& \ (\forall i \in dom(f))(f'(i) \geq_{n+1} f(i))$. Let $G = G_x$ be the $<_L$-least L_η-generic subset of $\mathbb{P}$. Set $b^x_i = \{f(i) \mid f \in G\}$, each $i \in \omega$. Clearly, each b^x_i is a branch of t_x, $i \neq j \rightarrow b^x_i \neq b^x_j$, and $t_x \subseteq \bigcup_{i \in \omega} b^x_i$. Set $T^{n+1}_\alpha = \{\bigcup b^x_i \mid x \in T^n_\alpha \ \& \ i \in \omega\}$.

Stage $\alpha+1$, $\lim(\alpha)$. For each $n \in \omega$, set $T^n_{\alpha+1} = \{x^\frown\langle i\rangle \mid x \in T^n_\alpha \ \& \ i \in \omega\}$. Let $x \in T^n_\alpha$, $i \in \omega$. We define $t_{x^\frown\langle i\rangle}$ as follows: $t_{x^\frown\langle i\rangle} = t_x \cup \{\cup b^x_{\langle i,j\rangle} \mid j \in \omega\}$, where b^x_k are as above. By the genericity of the construction of T^{n+1}_α (above), we clearly have $t_x \subseteq \cup_{j\in\omega} b^x_{\langle i,j\rangle}$ for each fixed $i \in \omega$, and $i \neq j \rightarrow t_{x^\frown\langle i\rangle} \cap t_{x^\frown\langle j\rangle} = t_x$.

That completes the construction. Clearly, T^0 will be a Souslin tree in L. Suppose b_0 is any branch of T^0 (necessarily L-generic, of course). Define $\hat{T}^1 \subseteq T^1$ in $L[b_0]$ by $\hat{T}^1 = \cup_{x\in b_0} t_x$. Note that as T^0 is Souslin in L, $\omega_1^{L[b_0]} = \omega_1^L$, so $\hat{T}^1$ is clearly an ω_1-tree in $L[b_0]$. We show that, in fact, $\hat{T}^1$ is Souslin in $L[b_0]$. Work in $L[b_0]$. Write T for $\hat{T}^1$. Assume $A \subseteq T$ is a maximal antichain. Let $M \prec L_{\omega_2}[b_0]$ be the smallest M such that $A, T \in M$. Set $\alpha = \omega_1 \cap M$. Let $\pi : M \cong L_\beta[b_0 \restriction \alpha]$. Then $\pi(\omega_1) = \alpha$, $\pi(T) = T|\alpha$, $\pi(A) = A \cap (T|\alpha)$, and $A \cap (T|\alpha)$ is a maximal antichain of $T|\alpha$ in $L_\beta[b_0 \restriction \alpha]$. Now, α is uncountable in $L_\beta[b_0 \restriction \alpha]$, hence also in L_β. But α is countable in $L_{\eta_{1,\alpha}}$. Hence $\beta < \eta_{1,\alpha}$. But look, $T^0|\alpha+1 \in L_{\eta_{1,\alpha}}$, so $b_0 \restriction \alpha \in L_{\eta_{1,\alpha}}$. Thus $L_\beta[b_0 \restriction \alpha] \subseteq L_{\eta_{1,\alpha}}$. But then $A \cap (T|\alpha)$ lies in $L_{\eta_{1,\alpha}}$, whence $A \cap (T|\alpha)$ is maximal in $T|\alpha+1$, and hence in T. Thus $A = A \cap (T|\alpha)$, and we are done. Similarly, if b_1 is a branch of $\hat{T}^1$, then $\hat{T}^2 = \cup_{x\in b_1} t_x$ is a Souslin tree in $L[b_0, b_1]$, etc.

Suppose now that $b_0, b_1, \dots$ is an arbitrary sequence of branches, b_0 a branch of T^0, b_1 a branch of $\hat{T}^1 = \cup_{x\in b_0} t_x$, etc. We show that $L[\langle b_0, b_1, \dots\rangle] = L[a]$ for some $a \subseteq \omega$.

In L, define a function f as follows: Given $n \in \omega$, $x \in T^{n+1}$, $ht(x) \geq 1$, let $f(x) =$ the T^n-least $y \in T^n$ such that $x \in t_y$. It is easily seen that f is well-defined, and that $ht(f(x)) = ht(x) + 1$ for all x.

Let $a = \{\langle n,i\rangle \mid \langle n,i\rangle \in b_n \cap T^n_1\}$. Clearly, $L[a] \subseteq L[\langle b_0, b_1 \dots\rangle]$. Conversely, we show that $L[\langle b_0, b_1, \dots\rangle] \subseteq L[a]$.

Working in $L[a]$, define a sequence $\langle b'_n | n < \omega \rangle$ of sets as follows:

(i) Let $x_0^n = \langle n \rangle$, each $n \in \omega$.

(ii) Let x_1^n = that pair $\langle n,i \rangle$ such that $\langle n,i \rangle \in a$, each $n \in \omega$.

(iii) Let $x_{\alpha+1}^n = f(x_\alpha^{n+1})$, each $n \in \omega$, whenever $\alpha < \omega_1$ and each x_α^{n+1} is defined.

(iv) Let x_α^n be the unique sucessor of $\{x_\beta^n \mid \beta < \alpha\}$ on T_α^n whenever $\{x_\beta^n \mid \beta < \alpha\}$ is a branch of $T^n|\alpha$ which extends on T_α^n, each $n \in \omega$, <u>when</u> $\alpha < \omega_1$, $\lim(\alpha)$, and each x_β^n, $\beta < \alpha$, is defined.

(v) Let $\gamma \leq \omega_1^L$ be the first point where the above definition breaks down, and set $b'_n = \{x_\alpha^n \mid \alpha < \gamma\}$.

But by a simple induction on α, $(\forall n \in \omega)(x_\alpha^n \in b_n)$. Hence $\gamma = \omega_1^L$, and $b'_n = b_n$ for each n. Thus $\langle b_0, b_1, \ldots \rangle \in L[a]$, and we are done.

REFERENCES

Ba 1 J. Ball: A Note on the Separability of an Ordered Space, Can. Journ. Math. 7 (1955), 548-551.

Bu 1 J. Baumgartner: Decompositions and Embedding of Trees, Notices Amer. Math. Soc. 17 (1970), 967.

De 1 K.J. Devlin: Aspects of Constructibility, Springer Lecture Notes 354 (1973).

De 2 K.J. Devlin: Note on a Theorem of J. Baumgartner, Fund. Math. 76 (1972), 255-260.

Do 1 C.H. Dowker: On Countably Paracompact Spaces, Can. Journ. Math. 3 (1951), 29-224.

Fe 1 U. Felgner: Models of ZF Set Theory, Springer Lecture Notes 223 (1971).

Ga Sp H. Gaifman and E.P. Specker: Isomorphism Types of Trees, Proc. Amer. Math. Soc. 15 (1964), 1-7.

Je 1 T.J. Jech: Lectures in Set Theory, Springer Lecture Notes 217 (1971).

Je 2 T.J. Jech: Trees, Journal Symb. Logic 36 (1971), 1-14.

Je 3 T.J. Jech: Automorphisms of ω_1-trees, Trans. Amer. Math. Soc. 173 (1972), 57-70.

Jn 1 R.B. Jensen: Souslin's Hypothesis is Incompatible with $V = L$, Notices Amer. Math. Soc. 15 (1968), 935.

Jn 2 R.B. Jensen: Automorphism Properties of Souslin continua, Notices Amer. Math. Soc. 16 (1969), 576.

Jn Jo R.B. Jensen and H. Johnsbråten: A New Construction of a Non-Constructible Δ^1_3 Subset of ω, Fund. Math., to appear.

Ku 1 G. Kurepa: Emsembles ordonnés et ramifiés, Publ. Math. Univ. Belgrade 4 (1935), 1-138.

Ma So D.A. Martin and R.M. Solovay: Internal Cohen Extensions, Annals of Math. Logic 2 (1970), 143-178.

Mi 1 E.W. Miller: A Note on Souslin's Problem, Amer. Journ. Math. 65 (1943), 673-678.

Ru 1 M.E. Rudin: Countable Paracompactness and Souslin's Problem, Can. Journ. Math. 7 (1955), 543-547.

Ru 2 M.E. Rudin: Souslin's Conjecture, Amer. Math. Monthly 76 (1969), 1113-1119.

Ru 3 M.E. Rudin: A Normal Space X for which $X \times I$ is not normal, Fund. Math. 73 (1971), 179-186.

Sh 1 J.R. Shoenfield: Unramified forcing, in Axiomatic Set Theory, Proceedings of Symposia in Pure Mathematics, Amer. Math. Soc., 357-381.

So Te R.M. Solovay and S. Tennenbaum: Iterated Cohen Extensions and Souslin's Problem, Annals of Math. 94 (1971), 201-245.

Su 1 M. Souslin: Problème 3, Fund. Math. 1 (1920), 223.

GLOSSARY OF NOTATION

SYMBOL	APPROXIMATE MEANING	PAGE
V	the universe	
V_α	the α'th level in the cumulative hierarchy	
dom(f)	domain of the function f	
$f \restriction A$	the restriction of f to A	
$\lvert X \rvert$	the cardinality of the set X	
$\mathcal{P}(X)$	the power set of X, i.e. $\{Y \mid Y \subseteq X\}$	
2^α	$\{f \mid f : \alpha \to 2\}$ or the cardinality of this set, according to context	
$2^{<\alpha}$	$\bigcup_{\beta<\alpha} 2^\beta$	
On	the set of all ordinals	
$\lim(\alpha)$	α is a limit ordinal	
$\mathbb{Q}$, $\mathbb{R}$	the rational and real numbers (as ordered sets)	
CH	the continuum hypothesis	
GCH	the generalised continuum hypothesis	
ZF^-	the theory ZF without the power set axiom	
$A \models \varphi$	A is a model of φ	
$f : A \Leftrightarrow B$	f is an isomorphism between A and B	
$A \cong B$	A is isomorphic to B	
$A \prec B$	A is an elementary submodel of B	
LST	the language of set theory	1
F"X	the "F-image" of X	1
Def(X)	the set of all "definable" subsets of X	1
L	the constructible universe	2
L_α	the α'th level in the constructible hierarchy	2
$V = L$	the axiom of constructibility	2
$<_L$	the canonical well-ordering of L	2

INDEX

Vol. 247: Lectures on Operator Algebras. Tulane University Ring and Operator Theory Year, 1970–1971. Volume II. XI, 786 pages. 1972. DM 40,–

Vol. 248: Lectures on the Applications of Sheaves to Ring Theory. Tulane University Ring and Operator Theory Year, 1970–1971. Volume III. VIII, 315 pages. 1971. DM 26,–

Vol. 249: Symposium on Algebraic Topology. Edited by P. J. Hilton. VII, 111 pages. 1971. DM 16,–

Vol. 250: B. Jónsson, Topics in Universal Algebra. VI, 220 pages. 1972. DM 20,–

Vol. 251: The Theory of Arithmetic Functions. Edited by A. A. Gioia and D. L. Goldsmith VI, 287 pages. 1972. DM 24,–

Vol. 252: D. A. Stone, Stratified Polyhedra. IX, 193 pages. 1972. DM 18,–

Vol. 253: V. Komkov, Optimal Control Theory for the Damping of Vibrations of Simple Elastic Systems. V, 240 pages. 1972. DM 20,–

Vol. 254: C. U. Jensen, Les Foncteurs Dérivés de $\varprojlim$ et leurs Applications en Theorie des Modules. V, 103 pages. 1972. DM 16,–

Vol. 255: Conference in Mathematical Logic – London '70. Edited by W. Hodges. VIII, 351 pages. 1972. DM 26,–

Vol. 256: C. A. Berenstein and M. A. Dostal, Analytically Uniform Spaces and their Applications to Convolution Equations. VII, 130 pages. 1972. DM 16,–

Vol. 257: R. B. Holmes, A Course on Optimization and Best Approximation. VIII, 233 pages. 1972. DM 20,–

Vol. 258: Séminaire de Probabilités VI. Edited by P. A. Meyer. VI, 253 pages. 1972. DM 22,–

Vol. 259: N. Moulis, Structures de Fredholm sur les Variétés Hilbertiennes. V, 123 pages. 1972. DM 16,–

Vol. 260: R. Godement and H. Jacquet, Zeta Functions of Simple Algebras. IX, 188 pages. 1972. DM 18,–

Vol. 261: A. Guichardet, Symmetric Hilbert Spaces and Related Topics. V, 197 pages. 1972. DM 18,–

Vol. 262: H. G. Zimmer, Computational Problems, Methods, and Results in Algebraic Number Theory. V, 103 pages. 1972. DM 16,–

Vol. 263: T. Parthasarathy, Selection Theorems and their Applications. VII, 101 pages. 1972. DM 16,–

Vol. 264: W. Messing, The Crystals Associated to Barsotti-Tate Groups: With Applications to Abelian Schemes. III, 190 pages. 1972. DM 18,–

Vol. 265: N. Saavedra Rivano, Catégories Tannakiennes. II, 418 pages. 1972. DM 26,–

Vol. 266: Conference on Harmonic Analysis. Edited by D. Gulick and R. L. Lipsman. VI, 323 pages. 1972. DM 24,–

Vol. 267: Numerische Lösung nichtlinearer partieller Differential- und Integro-Differentialgleichungen. Herausgegeben von R. Ansorge und W. Törnig, VI, 339 Seiten. 1972. DM 26,–

Vol. 268: C. G. Simader, On Dirichlet's Boundary Value Problem. IV, 238 pages. 1972. DM 20,–

Vol. 269: Théorie des Topos et Cohomologie Etale des Schémas. (SGA 4). Dirigé par M. Artin, A. Grothendieck et J. L. Verdier. XIX, 525 pages. 1972. DM 50,–

Vol. 270: Théorie des Topos et Cohomologie Etale des Schémas. Tome 2. (SGA 4). Dirigé par M. Artin, A. Grothendieck et J. L. Verdier. V, 418 pages. 1972. DM 50,–

Vol. 271: J. P. May, The Geometry of Iterated Loop Spaces. IX, 175 pages. 1972. DM 18,–

Vol. 272: K. R. Parthasarathy and K. Schmidt, Positive Definite Kernels, Continuous Tensor Products, and Central Limit Theorems of Probability Theory. VI, 107 pages. 1972. DM 16,–

Vol. 273: U. Seip, Kompakt erzeugte Vektorräume und Analysis. IX, 119 Seiten. 1972. DM 16,–

Vol. 274: Toposes, Algebraic Geometry and Logic. Edited by. F. W. Lawvere. VI, 189 pages. 1972. DM 18,–

Vol. 275: Séminaire Pierre Lelong (Analyse) Année 1970–1971. VI, 181 pages. 1972. DM 18,–

Vol. 276: A. Borel, Représentations de Groupes Localement Compacts. V, 98 pages. 1972. DM 16,–

Vol. 277: Séminaire Banach. Edité par C. Houzel. VII, 229 pages. 1972. DM 20,–

Vol. 278: H. Jacquet, Automorphic Forms on GL(2). Part II. XIII, 142 pages. 1972. DM 16,–

Vol. 279: R. Bott, S. Gitler and I. M. James, Lectures on Algebraic and Differential Topology. V, 174 pages. 1972. DM 18,–

Vol. 280: Conference on the Theory of Ordinary and Partial Differential Equations. Edited by W. N. Everitt and B. D. Sleeman. XV, 367 pages. 1972. DM 26,–

Vol. 281: Coherence in Categories. Edited by S. Mac Lane. VII, 235 pages. 1972. DM 20,–

Vol. 282: W. Klingenberg und P. Flaschel, Riemannsche Hilbertmannigfaltigkeiten. Periodische Geodätische. VII, 211 Seiten. 1972. DM 20,–

Vol. 283: L. Illusie, Complexe Cotangent et Déformations II. VII, 304 pages. 1972. DM 24,–

Vol. 284: P. A. Meyer, Martingales and Stochastic Integrals I. VI, 89 pages. 1972. DM 16,–

Vol. 285: P. de la Harpe, Classical Banach-Lie Algebras and Banach-Lie Groups of Operators in Hilbert Space. III, 160 pages. 1972. DM 16,–

Vol. 286: S. Murakami, On Automorphisms of Siegel Domains. V, 95 pages. 1972. DM 16,–

Vol. 287: Hyperfunctions and Pseudo-Differential Equations. Edited by H. Komatsu. VII, 529 pages. 1973. DM 36,–

Vol. 288: Groupes de Monodromie en Géométrie Algébrique. (SGA 7 I). Dirigé par A. Grothendieck. IX, 523 pages. 1972. DM 50,–

Vol. 289: B. Fuglede, Finely Harmonic Functions. III, 188. 1972. DM 18,–

Vol. 290: D. B. Zagier, Equivariant Pontrjagin Classes and Applications to Orbit Spaces. IX, 130 pages. 1972. DM 16,–

Vol. 291: P. Orlik, Seifert Manifolds. VIII, 155 pages. 1972. DM 16,–

Vol. 292: W. D. Wallis, A. P. Street and J. S. Wallis, Combinatorics: Room Squares, Sum-Free Sets, Hadamard Matrices. V, 508 pages. 1972. DM 50,–

Vol. 293: R. A. DeVore, The Approximation of Continuous Functions by Positive Linear Operators. VIII, 289 pages. 1972. DM 24,–

Vol. 294: Stability of Stochastic Dynamical Systems. Edited by R. F. Curtain. IX, 332 pages. 1972. DM 26,–

Vol. 295: C. Dellacherie, Ensembles Analytiques, Capacités, Mesures de Hausdorff. XII, 123 pages. 1972. DM 16,–

Vol. 296: Probability and Information Theory II. Edited by M. Behara, K. Krickeberg and J. Wolfowitz. V, 223 pages. 1973. DM 20,–

Vol. 297: J. Garnett, Analytic Capacity and Measure. IV, 138 pages. 1972. DM 16,–

Vol. 298: Proceedings of the Second Conference on Compact Transformation Groups. Part 1. XIII, 453 pages. 1972. DM 32,–

Vol. 299: Proceedings of the Second Conference on Compact Transformation Groups. Part 2. XIV, 327 pages. 1972. DM 26,–

Vol. 300: P. Eymard, Moyennes Invariantes et Représentations Unitaires. II. 113 pages. 1972. DM 16,–

Vol. 301: F. Pittnauer, Vorlesungen über asymptotische Reihen. VI, 186 Seiten. 1972. DM 18,–

Vol. 302: M. Demazure, Lectures on p-Divisible Groups. V, 98 pages. 1972. DM 16,–

Vol. 303: Graph Theory and Applications. Edited by Y. Alavi, D. R. Lick and A. T. White. IX, 329 pages. 1972. DM 26,–

Vol. 304: A. K. Bousfield and D. M. Kan, Homotopy Limits, Completions and Localizations. V, 348 pages. 1972. DM 26,–

Vol. 305: Théorie des Topos et Cohomologie Etale des Schémas. Tome 3. (SGA 4). Dirigé par M. Artin, A. Grothendieck et J. L. Verdier. VI, 640 pages. 1973. DM 50,–

Vol. 306: H. Luckhardt, Extensional Gödel Functional Interpretation. VI, 161 pages. 1973. DM 18,–

Vol. 307: J. L. Bretagnolle, S. D. Chatterji et P.-A. Meyer, Ecole d'été de Probabilités: Processus Stochastiques. VI, 198 pages. 1973. DM 20,–

Vol. 308: D. Knutson, λ-Rings and the Representation Theory of the Symmetric Group. IV, 203 pages. 1973. DM 20,–

Vol. 309: D. H. Sattinger, Topics in Stability and Bifurcation Theory. VI, 190 pages. 1973. DM 18,–

Vol. 310: B. Iversen, Generic Local Structure of the Morphisms in Commutative Algebra. IV, 108 pages. 1973. DM 16,–

Vol. 311: Conference on Commutative Algebra. Edited by J. W. Brewer and E. A. Rutter. VII, 251 pages. 1973. DM 22,–

Vol. 312: Symposium on Ordinary Differential Equations. Edited by W. A. Harris, Jr. and Y. Sibuya. VIII, 204 pages. 1973. DM 22,–

Vol. 313: K. Jörgens and J. Weidmann, Spectral Properties of Hamiltonian Operators. III, 140 pages. 1973. DM 16,–

Vol. 314: M. Deuring, Lectures on the Theory of Algebraic Functions of One Variable. VI, 151 pages. 1973. DM 16,–

Vol. 315: K. Bichteler, Integration Theory (with Special Attention to Vector Measures). VI, 357 pages. 1973. DM 26,–

Vol. 316: Symposium on Non-Well-Posed Problems and Logarithmic Convexity. Edited by R. J. Knops. V, 176 pages. 1973. DM 18,–

Vol. 317: Séminaire Bourbaki – vol. 1971/72. Exposés 400–417. IV, 361 pages. 1973. DM 26,–

Vol. 318: Recent Advances in Topological Dynamics. Edited by A. Beck, VIII, 285 pages. 1973. DM 24,–

Vol. 319: Conference on Group Theory. Edited by R. W. Gatterdam and K. W. Weston. V, 188 pages. 1973. DM 18,–

Vol. 320: Modular Functions of One Variable I. Edited by W. Kuyk. V, 195 pages. 1973. DM 18,–

Vol. 321: Séminaire de Probabilités VII. Edité par P. A. Meyer. VI, 322 pages. 1973. DM 26,–

Vol. 322: Nonlinear Problems in the Physical Sciences and Biology. Edited by I. Stakgold, D. D. Joseph and D. H. Sattinger. VIII, 357 pages. 1973. DM 26,–

Vol. 323: J. L. Lions, Perturbations Singulières dans les Problèmes aux Limites et en Contrôle Optimal. XII, 645 pages. 1973. DM 42,–

Vol. 324: K. Kreith, Oscillation Theory. VI, 109 pages. 1973. DM 16,–

Vol. 325: Ch.-Ch. Chou, La Transformation de Fourier Complexe et L'Equation de Convolution. IX, 137 pages. 1973. DM 16,–

Vol. 326: A. Robert, Elliptic Curves. VIII, 264 pages. 1973. DM 22,–

Vol. 327: E. Matlis, 1-Dimensional Cohen-Macaulay Rings. XII, 157 pages. 1973. DM 18,–

Vol. 328: J. R. Büchi and D. Siefkes, The Monadic Second Order Theory of All Countable Ordinals. VI, 217 pages. 1973. DM 20,–

Vol. 329: W. Trebels, Multipliers for (C, α)-Bounded Fourier Expansions in Banach Spaces and Approximation Theory. VII, 103 pages. 1973. DM 16,–

Vol. 330: Proceedings of the Second Japan-USSR Symposium on Probability Theory. Edited by G. Maruyama and Yu. V. Prokhorov. VI, 550 pages. 1973. DM 36,–

Vol. 331: Summer School on Topological Vector Spaces. Edited by L. Waelbroeck. VI, 226 pages. 1973. DM 20,–

Vol. 332: Séminaire Pierre Lelong (Analyse) Année 1971-1972. V, 131 pages. 1973. DM 16,–

Vol. 333: Numerische, insbesondere approximationstheoretische Behandlung von Funktionalgleichungen. Herausgegeben von R. Ansorge und W. Törnig. VI, 296 Seiten. 1973. DM 24,–

Vol. 334: F. Schweiger, The Metrical Theory of Jacobi-Perron Algorithm. V, 111 pages. 1973. DM 16,–

Vol. 335: H. Huck, R. Roitzsch, U. Simon, W. Vortisch, R. Walden, B. Wegner und W. Wendland, Beweismethoden der Differentialgeometrie im Großen. IX, 159 Seiten. 1973. DM 18,–

Vol. 336: L'Analyse Harmonique dans le Domaine Complexe. Edité par E. J. Akutowicz. VIII, 169 pages. 1973. DM 18,–

Vol. 337: Cambridge Summer School in Mathematical Logic. Edited by A. R. D. Mathias and H. Rogers. IX, 660 pages. 1973. DM 42,–

Vol: 338: J. Lindenstrauss and L. Tzafriri, Classical Banach Spaces. IX, 243 pages. 1973. DM 22,–

Vol. 339: G. Kempf, F. Knudsen, D. Mumford and B. Saint-Donat, Toroidal Embeddings I. VIII, 209 pages. 1973. DM 20,–

Vol. 340: Groupes de Monodromie en Géométrie Algébrique. (SGA 7 II). Par P. Deligne et N. Katz. X, 438 pages. 1973. DM 40,–

Vol. 341: Algebraic K-Theory I, Higher K-Theories. Edited by H. Bass. XV, 335 pages. 1973. DM 26,–

Vol. 342: Algebraic K-Theory II, "Classical" Algebraic K-Theory, and Connections with Arithmetic. Edited by H. Bass. XV, 527 pages. 1973. DM 36,–

Vol. 343: Algebraic K-Theory III, Hermitian K-Theory and Geometric Applications. Edited by H. Bass. XV, 572 pages. 1973. DM 38,–

Vol. 344: A. S. Troelstra (Editor), Metamathematical Investigation of Intuitionistic Arithmetic and Analysis. XVII, 485 pages. 1973. DM 34,–

Vol. 345: Proceedings of a Conference on Operator Theory. Edited by P. A. Fillmore. VI, 228 pages. 1973. DM 20,–

Vol. 346: Fučík et al., Spectral Analysis of Nonlinear Operators. II, 287 pages. 1973. DM 26,–

Vol. 347: J. M. Boardman and R. M. Vogt, Homotopy Invariant Algebraic Structures on Topological Spaces. X, 257 pages. 1973. DM 22,–

Vol. 348: A. M. Mathai and R. K. Saxena, Generalized Hypergeometric Functions with Applications in Statistics and Physical Sciences. VII, 314 pages. 1973. DM 26,–

Vol. 349: Modular Functions of One Variable II. Edited by W. Kuyk and P. Deligne. V, 598 pages. 1973. DM 38,–

Vol. 350: Modular Functions of One Variable III. Edited by W. Kuyk and J.-P. Serre. V, 350 pages. 1973. DM 26,–

Vol. 351: H. Tachikawa, Quasi-Frobenius Rings and Generalizations. XI, 172 pages. 1973. DM 18,–

Vol. 352: J. D. Fay, Theta Functions on Riemann Surfaces. V, 137 pages. 1973. DM 16,–

Vol. 353: Proceedings of the Conference. on Orders, Group Rings and Related Topics. Organized by J. S. Hsia, M. L. Madan and T. G. Ralley. X, 224 pages. 1973. DM 20,–

Vol. 354: K. J. Devlin, Aspects of Constructibility. XII, 240 pages. 1973. DM 22,–

Vol. 355: M. Sion, A Theory of Semigroup Valued Measures. V, 140 pages. 1973. DM 16,–

Vol. 356: W. L. J. van der Kallen, Infinitesimally Central-Extensions of Chevalley Groups. VII, 147 pages. 1973. DM 16,–

Vol. 357: W. Borho, P. Gabriel und R. Rentschler, Primideale in Einhüllenden auflösbarer Lie-Algebren. V, 182 Seiten. 1973. DM 18,–

Vol. 358: F. L. Williams, Tensor Products of Principal Series Representations. VI, 132 pages. 1973. DM 16,–

Vol. 359: U. Stammbach, Homology in Group Theory. VIII, 183 pages. 1973. DM 18,–

Vol. 360: W. J. Padgett and R. L. Taylor, Laws of Large Numbers for Normed Linear Spaces and Certain Fréchet Spaces. VI, 111 pages. 1973. DM 16,–

Vol. 361: J. W. Schutz, Foundations of Special Relativity: Kinematic Axioms for Minkowski Space Time. XX, 314 pages. 1973. DM 26,–

Vol. 362: Proceedings of the Conference on Numerical Solution of Ordinary Differential Equations. Edited by D. Bettis. VIII, 490 pages. 1974. DM 34,–

Vol. 363: Conference on the Numerical Solution of Differential Equations. Edited by G. A. Watson. IX, 221 pages. 1974. DM 20,–

Vol. 364: Proceedings on Infinite Dimensional Holomorphy. Edited by T. L. Hayden and T. J. Suffridge. VII, 212 pages. 1974. DM 20,–

Vol. 365: R. P. Gilbert, Constructive Methods for Elliptic Equations. VII, 397 pages. 1974. DM 26,–

Vol. 366: R. Steinberg, Conjugacy Classes in Algebraic Groups (Notes by V. V. Deodhar). VI, 159 pages. 1974. DM 18,–

Vol. 367: K. Langmann und W. Lütkebohmert, Cousinverteilungen und Fortsetzungssätze. VI, 151 Seiten. 1974. DM 16,–

Vol. 368: R. J. Milgram, Unstable Homotopy from the Stable Point of View. V, 109 pages. 1974. DM 16,–

Vol. 369: Victoria Symposium on Nonstandard Analysis. Edited by A. Hurd and P. Loeb. XVIII, 339 pages. 1974. DM 26,–

Vol. 370: B. Mazur and W. Messing, Universal Extensions and One Dimensional Crystalline Cohomology. VII, 134 pages. 1974. DM 16,–

Vol. 371: V. Poenaru, Analyse Différentielle. V, 228 pages. 1974. DM 20,-

Vol. 372: Proceedings of the Second International Conference on the Theory of Groups 1973. Edited by M. F. Newman. VII, 740 pages. 1974. DM 48,-

Vol. 373: A. E. R. Woodcock and T. Poston, A Geometrical Study of the Elementary Catastrophes. V, 257 pages. 1974. DM 22,-

Vol. 374: S. Yamamuro, Differential Calculus in Topological Linear Spaces. IV, 179 pages. 1974. DM 18,-

Vol. 375: Topology Conference 1973. Edited by R. F. Dickman Jr. and P. Fletcher. X, 283 pages. 1974. DM 24,-

Vol. 376: D. B. Osteyee and I. J. Good, Information, Weight of Evidence, the Singularity between Probability Measures and Signal Detection. XI, 156 pages. 1974. DM 16.-

Vol. 377: A. M. Fink, Almost Periodic Differential Equations. VIII, 336 pages. 1974. DM 26,-

Vol. 378: TOPO 72 - General Topology and its Applications. Proceedings 1972. Edited by R. Alò, R. W. Heath and J. Nagata. XIV, 651 pages. 1974. DM 50,-

Vol. 379: A. Badrikian et S. Chevet, Mesures Cylindriques, Espaces de Wiener et Fonctions Aléatoires Gaussiennes. X, 383 pages. 1974. DM 32,-

Vol. 380: M. Petrich, Rings- and Semigroups. VIII, 182 pages. 1974. DM 18,-

Vol. 381: Séminaire de Probabilités VIII. Edité par P. A. Meyer. IX, 354 pages. 1974. DM 32,-

Vol. 382: J. H. van Lint, Combinatorial Theory Seminar Eindhoven University of Technology. VI, 131 pages. 1974. DM 18,-

Vol. 383: Séminaire Bourbaki - vol. 1972/73. Exposés 418-435. IV, 334 pages. 1974. DM 30,-

Vol. 384: Functional Analysis and Applications, Proceedings 1972. Edited by L. Nachbin. V, 270 pages. 1974. DM 22,-

Vol. 385: J. Douglas Jr. and T. Dupont, Collocation Methods for Parabolic Equations in a Single Space Variable (Based on C^1-Piecewise-Polynomial Spaces). V, 147 pages. 1974. DM 16,-

Vol. 386: J. Tits, Buildings of Spherical Type and Finite BN-Pairs. IX, 299 pages. 1974. DM 24,-

Vol. 387: C. P. Bruter, Eléments de la Théorie des Matroïdes. V, 138 pages. 1974. DM 18,-

Vol. 388: R. L. Lipsman, Group Representations. X, 166 pages. 1974. DM 20,-

Vol. 389: M.-A. Knus et M. Ojanguren, Théorie de la Descente et Algèbres d' Azumaya. IV, 163 pages. 1974. DM 20,-

Vol. 390: P. A. Meyer, P. Priouret et F. Spitzer, Ecole d'Eté de Probabilités de Saint-Flour III - 1973. Edité par A. Badrikian et P.-L. Hennequin. VIII, 189 pages. 1974. DM 20,-

Vol. 391: J. Gray, Formal Category Theory: Adjointness for 2-Categories. XII, 282 pages. 1974. DM 24,-

Vol. 392: Géométrie Différentielle, Colloque, Santiago de Compostela, Espagne 1972. Edité par E. Vidal. VI, 225 pages. 1974. DM 20,-

Vol. 393: G. Wassermann, Stability of Unfoldings. IX, 164 pages. 1974. DM 20,-

Vol. 394: W. M. Patterson 3rd, Iterative Methods for the Solution of a Linear Operator Equation in Hilbert Space - A Survey. III, 183 pages. 1974. DM 20,-

Vol. 395: Numerische Behandlung nichtlinearer Integrodifferential- und Differentialgleichungen. Tagung 1973. Herausgegeben von R. Ansorge und W. Törnig. VII, 313 Seiten. 1974. DM 28,-

Vol. 396: K. H. Hofmann, M. Mislove and A. Stralka, The Pontryagin Duality of Compact O-Dimensional Semilattices and its Applications. XVI, 122 pages. 1974. DM 18,-

Vol. 397: T. Yamada, The Schur Subgroup of the Brauer Group. V, 159 pages. 1974. DM 18,-

Vol. 398: Théories de l'Information, Actes des Rencontres de Marseille-Luminy, 1973. Edité par J. Kampé de Fériet et C. Picard. XII, 201 pages. 1974. DM 23,-

Vol. 399: Functional Analysis and its Applications, Preceedings 1973. Edited by H. G. Garnir, K. R. Unni and J. H. Williamson. XVII, 569 pages. 1974. DM 44,-

Vol. 400: A Crash Course on Kleinian Groups - San Francisco 1974. Edited by L. Bers and I. Kra. VII, 130 pages. 1974. DM 18,-

Vol. 401: F. Atiyah, Elliptic Operators and Compact Groups. V, 93 pages. 1974. DM 18,-

Vol. 402: M. Waldschmidt, Nombres Transcendants. VIII, 277 pages. 1974. DM 25,-

Vol. 403: Combinatorial Mathematics - Proceedings 1972. Edited by D. A. Holton. VIII, 148 pages. 1974. DM 18,-

Vol. 404: Théorie du Potentiel et Analyse Hormonique. Edité par J. Faraut. V, 245 pages. 1974. DM 25,-

Vol. 405: K. Devlin and H. Johnsbråten, The Souslin Problem. VIII, 132 pages. 1974. DM 18,-